BREEDING EXOTIC BIRDS

A Beginner's Guide

First Edition

by

Fran Gonzalez

Published by
Neon Pet Publications
P.O. Box 733
Cypress, California 90630

Library of Congress Catalog Card Number: 93-93561

ISBN 0-9637844-0-4

Printed in the United States

Table of Contents

Chapter 4: Introducing Your Birds

Chapter 5: Feeding Your Birds

Chapter 6: Cages And Nest Boxes

Chapter 7: Eggs

Chapter 7: Eggs, Continued

Chapter 8: Care And Feeding Of The Babies

Chapter 9: Problems of Bird Breeding

Chapter 10: Illness

Chapter 11: Problems With Babies

INTRODUCTION

Breeding exotic birds can be very frustrating, especially when you are just getting started. You may talk to ten different people and get ten different answers to one simple question. Or, you may find that some successful breeders are unwilling to share their "secrets" with you. Many feel they learned the hard way and so should you. Fortunately, this pattern is changing, and more and more people are willingly sharing desperately needed tips and valuable information.

This book is designed for those of you who would like to begin breeding exotic birds and for those who have been in the hobby a short time.

This book will address all aspects involved in successful bird breeding: kinds of birds to start with, how to determine if birds are healthy, types of cages and nest boxes to use, where to position cages and nest boxes, how to check on eggs, when to pull babies, how to hand-feed, and how to wean babies. Additional sections will address incubating eggs and dealing with illness. A glossary of commonly used terms is located at the back of the book.

The techniques described in this book are those that we use with our birds. My husband, Omar, and I have been breeding exotic birds since 1982, and we have coached many people new to bird breeding and have helped them set up new pairs, using our methods. Many of those who followed our instructions have been rewarded with healthy baby birds.

Successful breeding is more than having a good pair of birds, it's *how* you set them up and care for them that determines how easily they will breed for you. Often new bird breeders talk to too many different people, using different methods. Trying to use all these different ideas *together* can cause one big, unsuccessful, confusing mess.

We encourage you to try our methods with your birds. Try them for at least a year. After that, follow some of the tips in Chapter 4, which deals with birds that show no inclination to breed.

This book will provide you with the necessary guidelines to begin breeding exotic birds; *try to stick to these guidelines as closely as possible.* The purpose of this book is to help you establish a system that works for you and your birds. Once your birds are settled in their breeding cages, and you become more familiar with them and their nesting habits, you may decide to experiment with new techniques, one at a time.

Our breeding setup is designed for cage-breeding exotic birds, rather than aviary (large flight) breeding, because we live in a fairly heavily populated urban area, and we cage breed our birds. It's easy to convert a garage, basement or extra room into a bird breeding facility. You can build a shed or attach cages to the side of your house, as long as you live in a mild climate. The possibilities are practically endless.

If you want to breed exotic birds for profit, plan on spending a lot of money the first year or two to develop a successful system. Keep in mind, however, that you will breed exotic birds mostly for pleasure. After all of the time and effort you will spend on your birds, you will probably make a minimal profit. Usually the money earned is spent on more birds!

How to Catch and Hold a Bird

Most people who buy exotic birds never learn how to properly catch and hold the birds they buy. It is extremely important for you to know how, especially in case of an emergency. Fixing a broken toenail or bloodfeather can be a traumatic experience for the bird, but even more so for the owner, if he or she doesn't know how to properly restrain the bird for an examination. Familiarize yourself with these photos, which show how two people would hold a bird, as well as by yourself.

First: Corner the bird so that it cannot fly around the room (or cage).

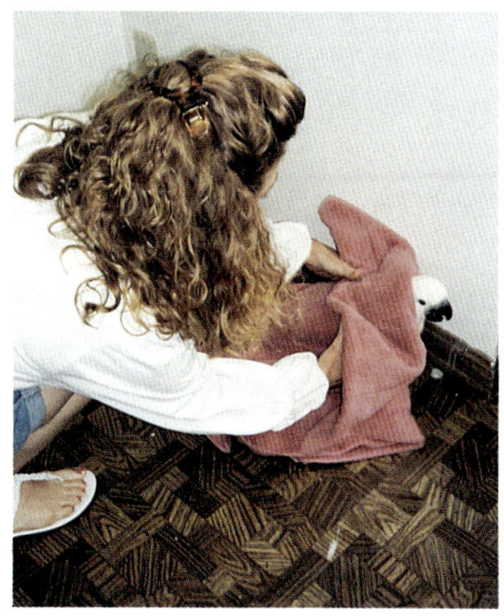

Second: Quickly drop a towel over the bird's entire body, paying attention to the bird's head.

Third: Grasp the bird around the neck and gather the towel around the bird's feet, keeping the bird snug inside the towel so it cannot wriggle free.

Fourth: To minimize the stress to the bird, work quickly, and try not to restrain the bird for more than five minutes.

To restrain a bird by yourself, catch the bird in a corner as described on the previous page then:

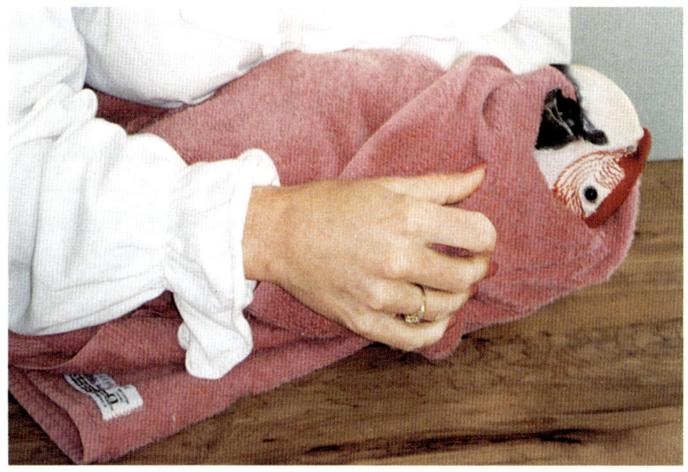

Tuck the bird snugly in the towel and only expose the bird's head.
DO NOT PRESS DOWN HARD ON THE BIRD'S CHEST.

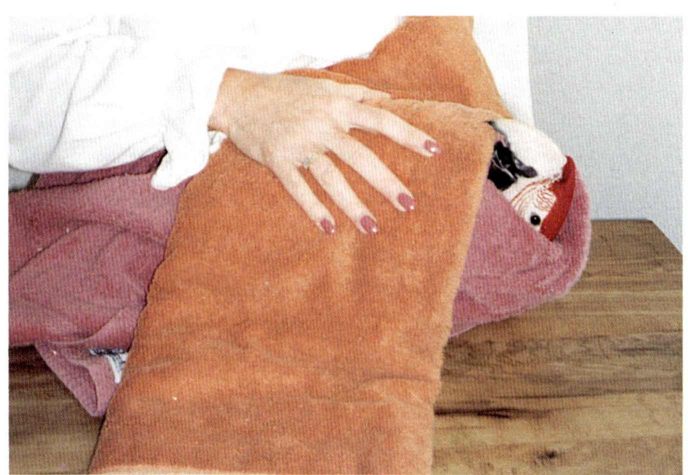

If necessary, place another towel on top of the bird to facilitate examining or clipping.

This technique is tricky at first; it works best if you practice on small birds, such as Cockatiels, or on tame birds. The most important things to remember when catching a bird are to be quick and try to stress the bird as little as possible.

CHAPTER 1

HOW TO CHOOSE BREEDER BIRDS

What Kinds of Birds Should You Start With?

If you have been breeding Cockatiels, Conures or other small birds, don't rush to sell all of them (some of which may be proven breeders) to breed large hookbills. The experience you gain with your smaller birds will be helpful, and any offspring they produce may help provide money to feed your newer, larger pairs, which are considerably more expensive to keep!

When choosing birds you must consider:

- Noise
- Size
- How easily they breed
- Cost
- How easily the babies will sell
- What birds appeal to you

Space

If you have lots of room and wish to convert a large room into a bird room then size is not a problem.

If you are limited in space, rather than using any available space on a few large pairs, consider setting up several small to medium size pairs.

Which Birds Are Easier to Breed?

It is always good to start with a pair of birds that will breed right away. Certain birds tend to breed easier than others. The following list may help you choose:

Difficult:

Caiques
Hawk-headed Parrots
Goffin's Cockatoos
Moluccan Cockatoos
Hyacinth Macaws
Green-winged Macaws

Somewhat Difficult:

Timneh African Greys
Blue-headed Pionus
Amazons
Lesser Sulphur-crested Cockatoos
Scarlet Macaws
Jardine's Parrots
Derbyan Parakeets

Easier:

Congo African Greys
Senegal Parrots
Blue and Gold Macaws
Eclectus Parrots
Umbrella Cockatoos
White-Capped Pionus
Mini-macaws
Conures
Quaker Parakeets
Ringneck Parakeets
Alexandrine Parakeets

Important Note: This chart is meant to give a general idea how readily certain bird species breed; every pair of birds is different, however.

Noise

Cockatoos, Macaws, Conures, Mini-macaws and Amazon Parrots are all very noisy.

Eclectus, Pionus, African Parrots and Caiques are less noisy.

AVERAGE NOISE LEVELS

LOW (Natural Noises)	MEDIUM (Noisy at times)	HIGH (Very Noisy)
African Parrots	Eclectus Parrots	Amazon Parrots
Caiques	Hawk-headed Parrots	Large Cockatoos
Pionus Parrots	Most Mini-macaws	Most Conures
Pyrrhura Conures	Small Cockatoos	Most Large Macaws
Ringneck Parakeets	Some Large Macaws	Severe Macaws

Remember: This chart rates species in general; individual pairs of birds may be different.

Cost

First, decide how much money you want to spend and don't get too carried away. Never take out a second mortgage or borrow a lot of money to buy birds, cages and equipment. Start with whatever you can afford and leave a little room in case an emergency arises.

Finally, the most important note of all: breed the birds *you* like. If you specialize in certain breeds, having more than one pair of the same type of bird stimulates the birds and encourages breeding activity for the group. You can then also hold back babies from unrelated pairs and create new pairs from your breeder's offspring!

CHOOSING A HEALTHY BIRD

How do you know if a bird is healthy? You don't, unless you take the bird to an avian veterinarian and have a thorough exam, which includes cultures and blood work. Lab work will be discussed in Chapter 2; for now we will discuss what things to look for when you buy a bird.

Feather Quality

The plumage, or feathers, should be clean and smooth. Feathers should not be broken, paler than normal, or contain stress lines.

Pale feathers could mean a vitamin deficiency, as can too many yellow feathers, especially those that grow on parts of the body that are not normally yellow. Yellow feathers can also be an indication of a liver disfunction or even overexposure to the sun.

Some birds chew their feathers or pluck them out; this is usually a sign of sexual frustration in a mature bird and *in no way* affects that bird's ability to breed. Such birds usually go to nest right away.

You must be careful not to confuse feather chewing or plucking with feather *loss*, which is much more serious. Feather loss can be caused by a thyroid imbalance or Psittacine Beak and Feather Disease (PBFD), also referred to as Cockatoo Feather Syndrome (CFS) or Cockatoo Rot.

Cockatoo Feather Syndrome is very common among recently imported Cockatoos. Typical symptoms include loss of powder; crest feathers pinched off and/or broken with black marks on them; body feathers that appear pinched in the middle and never develop; an overgrown beak; or a beak that chips easily.

Avoid buying Cockatoos right out of quarantine unless you feel experienced enough to choose a healthy bird. There are never any guarantees. There are many semi-tame and tame pet cockatoos available, and such birds may be a better bet than wild-caught birds recently released from quarantine. (At the time of this writing the importation of exotic birds is quickly declining, thus the demand for pet birds as breeders will increase.)

Important Note: Whatever type of bird you choose to breed, a mature, previously owned pet bird is almost always the best choice and will probably breed right away.

Nostrils / Eyes / Mouth / Beak

Always examine the nose, eyes and mouth of any potential purchase.

Nostrils should be clean and clear. Both nostril openings should be the same size. If one nostril is quite a bit larger than the other, this may indicate a past or present bacterial infection. There should be no discharge or wetness around the nose or on the beak.

Eyes should be clear and bright. Dull, glazed eyes could be a sign of illness. Both eyes should respond to rapid movement. Some birds will have one bad eye, and unless you look closely, it may seem almost as good as a perfect eye. If there is an opaque film across its eye, the bird is probably blind in that eye. Many partially sighted birds breed, but they should cost less money than a bird with no deformities.

Sometimes a bird will have an unusually shaped pupil (the black center part of the eye), or there will be a black speck on some part of the iris (the colored part of the eye). Such birds usually see well, but you should check their response to movement or have an avian veterinarian perform a vision exam.

Breeding Handicapped Birds or Those With Deformities

• Birds that are born handicapped or with deformities should not be bred; they may pass their same genetic problems to any offspring.

• Birds that are related should not be bred because it increases the chances of birth defects and weakens the gene pool.

• Birds that have lost toenails or toes in quarantine or from an accident make fine breeders, but should cost less than a perfect bird.

• Birds that are blind in one eye and birds that have an old break in a wing or a leg usually breed well, also.

• Birds with permanently damaged beaks can usually feed their mates and their babies, providing the damage is not extensive. If a bird with a beak deformity can feed itself, then it will probably be able to feed its babies.

CHAPTER 2

THE IMPORTANCE OF PROPER VETERINARY CARE

Once you've got an apparently healthy pair of birds, you can set them up with a cage and nest box, right? Wrong! It is very important to take the newly purchased birds to an avian veterinarian and make sure they are truly healthy.

Too many people skip this step either because they don't realize it is so essential, or they don't want to spend the money. This is a big mistake, because you can waste valuable time. Your pair of birds may lay eggs that are infertile, or they may lay fertile eggs that don't hatch. Or, even more frustrating, they may have babies that die soon after hatching. Then you have to try to figure out what went wrong.

It is much easier to treat a bird that has not become accustomed to a mate, nest box and breeding cage. If a pair of birds is found to be sick, nest box and perches must be thrown away and replaced, and the cage must be completely disinfected. It is much less stressful on you *and* the birds if you take care of this in the beginning, when you still have the birds in temporary holding cages.

Once you have decided you are going to take the bird in to the veterinarian, which tests are necessary? There are so many! Before discussing what tests should be done, let's go over the *basic* tests, how they work, and how accurate they are.

Physical Exam

During a physical exam, the veterinarian looks at the bird's overall appearance, feather quality, weight and mouth. During this exam, the veterinarian should:

- Weigh your bird.
- Look in your bird's mouth.
- Look at your bird's abdomen and vent.
- Listen to your bird's heart, lungs and air sacs.
- Recommend health-screening tests, if needed.

Birds have gram-positive and gram-negative bacteria present in their systems. Gram-positive bacteria is usually acceptable in a bird, and does not usually cause a bird to become ill. A light to moderate amount of gram-negative bacteria is usually acceptable, unless it is a potentially dangerous bacteria such as *Salmonella*, *Psuedomonas* or

Klebsiella, but a heavy or an abundant amount is not acceptable. Always ask your veterinarian what type of bacteria your bird has and if it is gram-positive or gram-negative.

When a bird shows signs of illness such as a runny nose, fluffed feathers, runny stools, weight loss or lethargy, it is *clinically ill*.

Some perfectly healthy *looking* birds will culture out with a heavy bacterial infection; this in known as a *sub-clinical* infection. In the wild, birds hide their illnesses to avoid being abandoned by their flock and attracting predators. In captivity, birds are known for this same behavior. Stress can reduce a bird's natural ability to fight a bacteria or a virus. The following situations may stress a bird:

- Bathing a bird improperly
- Changing its diet
- Moving its cage
- A sudden change in temperature
- Changing its routine
- Grooming
- A health exam

Gram's Stain

A Gram's stain is a simple procedure in which the veterinarian swabs a surface, usually the throat (called the choanal slit), or the bird's vent, a stool sample, nasal discharge, an unhatched egg or a baby.

The veterinarian then smears one or more glass slides with the swab and follows a procedure in which the slide is flooded with several solutions and rinsed accordingly. The end result is a slide the veterinarian can look at under the microscope to view the possible ratio of gram-positive vs. gram-negative bacteria. If the Gram's stain reflects normal levels of bacterial flora in a healthy-looking bird, no other testing should be necessary. If the Gram's stain appears to indicate a bacterial infection, and the bird is not clinically ill, it is best to do a culture and sensitivity instead of guessing about which antibiotics would be most effective. Of course, if the bird displays signs of illness, antibiotics could save its life.

We do not use the Gram's stain on any of our birds; we prefer to have our veterinarian do a complete culture and sensitivity.

Culture and Sensitivity

To more accurately check your bird's bacterial levels, you can have the veterinarian do a culture and sensitivity. The veterinarian swabs the roof of the bird's mouth (or other site) and then wipes the swab on a prepared bacteria culture plate. The plate is labeled and sealed and sent to a laboratory. At the lab, they put the plate in an incubator and see what type of bacteria grows. After about 48-72 hours, the lab identifies the types of bacteria present and determines if the growth is light, moderate or heavy.

They then test the bacteria with antibiotics to see which antibiotics are most effective. Many types of bacteria resist antibiotics. This is why it is important not to indiscriminately give antibiotics to your birds; it's too easy to give a bird an antibiotic that may not be effective for that bird's illness. A correct antibiotic at the wrong dose can cause the bacteria to develop a resistance to it.

A culture and sensitivity usually takes about 3-4 days. Many veterinarians now do their own cultures, which can speed up the process. Request a copy of the lab report so you can familiarize yourself with the test, its results and the drugs that will be effective against it.

We always have a choanal (throat) culture done on any new birds purchased for breeding. Commonly, birds have mild bacterial infections, which can easily be treated with antibiotics in the birds' drinking water. More serious infections may require oral or injectable treatments. If a bird cultures out with low levels of gram-negative bacteria but is otherwise healthy, we do not treat them with antibiotics. If we cultured our own mouths, they would be loaded with bacteria. Similarly, bird's mouths contain natural, bacterial flora.

Fecal Exam

It's also important to take a fresh stool sample to your veterinarian and have him/her examine it for internal parasites (worms). Imported birds are more likely to have parasites than birds that have been domestically raised or kept as pets for several years. Stool samples can be looked at under a microscope or checked by a flotation test for parasite eggs or worms. There are no false positives with this test, but birds can act as carriers, causing occasional false negatives.

Avian Blood Tests

Depending upon how much information you want and how much money you want to spend, you can learn a lot from your bird's blood profile.

Most veterinarians perform a simple CBC, or complete blood count, when they examine healthy-looking birds. This test checks the white and red cell count. An elevated white-cell count may indicate an infection, and you may wish to have the veterinarian do further testing.

Important Note: A slightly elevated white count can be caused by the stress of the visit to the veterinarian, so if all the other tests check out okay, the bird is probably fine. Be sure the veterinarian or technician doesn't "milk" the bird's toe when he or she takes the blood sample; it can affect the results. Some veterinarians prefer to take blood from the jugular or wing veins.

Full Avian Panel

A full avian blood panel will tell you everything from the red and white blood-cell count to the bird's calcium level, liver enzymes, uric acid and more. This is one of the most complete tests you can do on birds. It is also more expensive. African greys are prone to have calcium deficiencies, so this may be a test you wish to consider for a newly purchased bird of that species.

Psittacosis Testing

Testing for Psittacosis is very important. Psittacosis, or *Chlamydia psittaci,* is very common among imported birds and is easily treated if detected early. Psittacosis can be transmitted from one bird to another bird and from birds to humans, but it cannot be transmitted from human to human or human to bird. (In humans, the symptoms resemble a very bad respiratory flu that won't go away.)

Birds can carry Psittacosis for years, without appearing ill. Sometimes sick birds will exhibit bright, lime-green stools; others will show no signs at all.

Many methods are used to test for Psittacosis; the most common at this time are the Clamydia Antigen Test, and the Psittacosis Antibody Test.

Psittacosis can be a very tricky organism to test for, and false negatives as well as false positives are common. The antibody test, which is a blood titer, is the method we prefer. A blood sample is sent out and the number of antibodies present indicates if the bird could possibly have been exposed to, or be a carrier of, Psittacosis. A high number of antibodies could mean that your bird was recently exposed, and your veterinarian may recommend you have the bird re-tested in 30 days, providing the bird is acting fine and is otherwise healthy.

If the bird in question is clinically ill, a Clamydia Antigen, or swab test, is quicker and easier. Many avian veterinarians do this test in their own office, and even if they don't, the results are available within hours. The only drawback to this test is that it can give a false negative reading if the bird is not shedding the organism at the time of the test, even though the bird may be a carrier of Psittacosis. A bird that is shedding the organism is contagious. The disease could spread to other birds or to humans at that time.

We have all of our breeders checked for Psittacosis with the blood titer (antibody) method and if the titer comes back high, we test with the swab (antigen) test. If the test comes back positive, the bird should be treated with tetracycline therapy for 45 days.

There are other methods used to test for Psittacosis however, you may wish to consult with your veterinarian regarding how accurate these methods are, and what he or she recommends in each individual case.

Psittacine Beak and Feather Disease Testing

At the time of this writing, there are several methods available to test for Psittacine Beak and Feather Disease, or PBFD. Previous methods usually consisted of testing a feather sample of the bird. The latest test is a DNA probe test which is supposed to be more accurate. If a bird tests positive with the DNA method, it should be retested in 90 days. I feel that the testing methods will continue to improve, but there will still be many false positives.

Polyoma Virus Testing

Polyomavirus (Papovavirus) testing became available in 1992 through a swab test that can be performed on a bird or any surface the bird may come in contact with. This test is helpful for aviculturists trying to track the virus in their aviary, nursery, or incubator.

Papillomas and Prolapses

If you are getting into breeding, you must become familiar with Papillomas. A Papilloma is a cauliflower-like growth, usually inside the bird's chloaca, or vent. Some birds will also have Papilloma inside of their mouth. Papillomas also occur in humans. Papillomas are not always easy to see and are equally hard to cure. One current theory holds that a bird with Papillomas cannot breed successfully, however, I'm sure there are quite a few birds out there with papillomas that breed.

The best way to check for Papillomas is to gently insert a sterile swab into the birds vent and gently fold back the skin. (You may use a sterile lubricant.) If the area is clean and there are no growths, that is a good sign. A bird that has had Papilloma for a long time may appear to have what looks like dried fecal material around or below the vent. This dried matter is actually old Papilloma tissue.

Papillomas can be surgically removed, but they can grow back. Avian veterinarians are experimenting with serums made from the bird's own Papilloma tissue. At this time, this does not seem to be an effective method of treatment. There has been some success with the use of laser surgery, but unfortunately at this time the equipment is very expensive and not always available.

It is important not to confuse a Papilloma with a Prolapse. To an untrained eye a Prolapse looks almost identical to a Papilloma: A Prolapse looks like a hemorrhoid and may appear only when the bird is defecating or stressed. Sometimes a bird with a Prolapse has a serious bacterial infection, and the Prolapse will disappear after antibiotic therapy. In severe cases, however, your veterinarian may need to clean and suture (called a "purse-string") the weakened tissue. If the bird with the Prolapse is a female, she may have a severe calcium deficiency. If that is the case, calcium therapy is recommended. If you are hand-feeding a baby and notice a Prolapse, it may be the result of irritation from feeding the formula too thick or an irritation from the bedding material (common with wood shavings). It can also be parasites or a bacterial infection. You should take the bird to your veterinarian for an examination.

As you can see, there are lots of standard tests you can have done to see how healthy your bird is. NEVER euthanize a bird unless you are completely sure that it is incurably ill. Get a second opinion any time you have the slightest doubt about a bird. It is not unusual to experience false positives when testing exotic birds, and if the bird appears healthy, it deserves the time and expense of a definite diagnosis.

When we set up birds purchased from quarantine, we usually have our avian veterinarian do a throat culture, Psittacosis titer, fecal exam and mini blood panel on *each* bird. When we buy a bird that has been an ex-pet or was domestically raised, we usually have the veterinarian do only a throat culture and a Psittacosis titer.

CHAPTER 3

SEXING

There are several methods available to determine the sex of birds. If you plan to set up birds that will breed for you, it is essential that you have them *surgically sexed.* Surgical sexing, when done correctly, is a fast way to determine the sex of birds *and* to check the health of the internal reproductive systems.

Surgical Sexing

The bird is anesthetized with isoflurane gas and laid on its right side (female birds have only one functioning ovary on the left side, so the procedure must be done on the left side). A few feathers are plucked around the thigh area, and then the skin is swabbed with Betadine® solution, to kill bacteria and prevent infection.

A small incision is made and the veterinarian inserts a laparoscope, which he or she maneuvers past the air sacs to the reproductive organs. In mature birds, a male's testes looks like a navy bean, while a female's ovaries look like a bunch of grapes. In immature birds, it is sometimes difficult to determine the sex, because the organs look very similar. Many times the veterinarian can tell how mature the bird is by the shape and size of the reproductive organs. After that, the veterinarian removes the scope and may smear more Betadine® on the incision. If there is some bleeding, it is usually stopped with pressure applied to the site. Some veterinarians also use one stitch.

The attending veterinarian who does the sexing should "tattoo" the bird to identify it as a male or female. This is done with a non toxic ink injected under the skin of the bird's right or left wing. Males are tattooed under the right wing; females on the left. Bird breeders jokingly remember this by the phrase, "Men are always right."

Next, the anesthesia is turned off, and the bird is put in a carrier or cage for a short recovery period. At this time, isoflurane gas is the safest and least harmful anesthesia for birds. Using injectable anesthesia in any procedure increases the risk of losing a bird and makes the recovery period longer and more stressful. When birds wake up from the gas, they can eat, drink and perch within 15 minutes or less. When a bird wakes up from injectable anesthesia it will often thrash around, so it is important that the bird is properly restrained (wrapped in a towel, or several layers of newspaper) so it cannot injure itself.

It is not uncommon, especially in African Greys, for birds to bloat with air after surgical sexing. This is due to a ruptured air sac, and the condition will often correct itself. In severe cases, it is necessary to have a veterinarian withdraw the air with a needle.

When choosing a veterinarian for surgical sexing, make sure he or she has the right equipment and does the procedure regularly. Some veterinarians have equipment to video tape the sexing and will let you watch during or after the procedure. There is still a slight risk of losing a bird during surgical sexing, however, this is mostly due to complications with bleeding, or the anesthesia, not the procedure itself. Surgical sexing should not be practiced without anesthesia! Most birds that die during or after surgical sexing are found to have pre-existing internal medical problems.

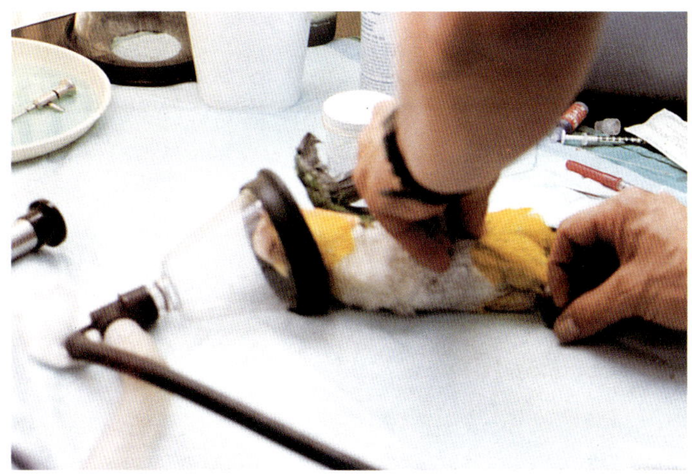

Surgical sexing is still the method many bird breeders choose, as it gives the breeder an indication of how mature the bird may be. This White-bellied Caique is ready for sexing.

Re-sexing a Bird

If you buy a bird from a dealer, ask the dealer to guarantee the sex he or she claims the bird to be. Some people will not do this, and you must then decide if it's worth taking a chance. Find out if the dealer knows when the bird was sexed and by whom. If there is less information than that to go by, *definitely* have the bird re-sexed! Birds should not be re-sexed sooner than one month after the original procedure because it is too stressful for the bird.

If you have a pair of birds that have been consistently laying infertile or clear eggs, you may wish to have the male re-sexed to see if his reproductive organs look normal.

Occasionally, a veterinarian may attempt to sex a bird but may not be able to clearly see the reproductive organs (especially if the bird is obese) or there is excessive bleeding. In such cases, it is best to stop and re-sex the bird at a later date.

Other Sexing Methods

The most reliable alternative to surgical sexing at this time is *DNA sexing*: For this method, a veterinarian takes a small blood sample (only one tube, which is about three drops of blood) and sends it out to a lab for confirmation of the sex by matching it with DNA. DNA sexing is an alternative to re-sexing a bird when you only want to confirm the bird's sex.

Sexing birds by stool samples or feathers may not be as accurate in determining a bird's sex, and you also do not have the opportunity to get an inside view of the bird's reproductive organs.

It is extremely important that you know for sure that you have a male and a female when you set up a new pair of birds, otherwise you may waste a lot of time.

In some species, you can determine the gender by looking at the bird. In such cases, you must decide if you want to have a veterinarian perform a laparoscopy to check the health of the bird's reproductive organs. We usually choose not to have these birds surgically sexed, but may decide to have them DNA sexed, unless we have reason to believe their reproductive organs may not be healthy.

The following chart represents many popular birds and features that will help you determine their sex.

DETERMINING SEX BY PHYSICAL CHARACTERISTICS

TYPE OF BIRD	DISTINGUISHING FEATURES
Congo African Greys	A male is usually a bit larger, has a square head, wider neck and an overall darker gray color than a female. A female is sometimes smaller, lighter in color, has a rounder head with longer neck than a male. A female's eyes are sometimes almond shaped, and she has gray tips on the red feathers under her tail. (These features are most accurate on a bird over two years of age.)

Eclectus Parrots (all sub-species)	A male is green. A female is red. (This is called sexual dimorphism)
White Fronted Amazon Parrots	A male has red primary feathers on his wings.
Most Amazon Parrots	A male is usually larger than a female, with a large, square head, and he is usually more aggressive than a female.
Senegal Parrots	A male sometimes has green feathers near his vent. His head will be large and square. Sometimes he will have more yellow on his breast. A female may have a longer green "V" on her chest than a male.
Scarlet Macaws	A male usually has a longer, more narrow beak than a female. A female has a shorter, wider beak (at the base of the mandible) than a male.
Most White Cockatoos, Tritons, Citrons, Sulphur-crested Cockatoos and Umbrella Cockatoos	When mature (after age 3), most females have reddish-brown eyes, while males' eyes are dark brown and appear black. This method is not reliable for Moluccans, Bare-eyeds, or Goffin's Cockatoos.
Moluccan and Goffin's Cockatoos	Males usually have larger beaks and are larger overall than females.
Alexandrine Parakeets, Derbyans, and Ringneck Parakeets	When mature (after about 1 year), A male develops a definite dark ring around his neck.

This Senegal Parrot has a short green "V" on its chest, which sometimes indicates a male.

CHAPTER 4

INTRODUCING YOUR BIRDS

Now that you have healthy, veterinarian checked birds, how do you introduce them?

Pairing Tame and Wild Birds

It is important to understand that the more tame a bird is, the more difficult it will be to pair it with another bird. This will come as a shock for many people, because there is a long-spread rumor that wild or untamed birds do not breed easily; this couldn't be further from the truth.

Some tame birds have been so domesticated (spoiled) and heavily imprinted on humans that they want nothing to do with other birds. They will even go out of their way to beat up another bird, intimidating the other bird so that any chance of the two breeding is slim.

Two wild, sexually mature birds will almost always seek security and companionship together and, if provided with the right environment, readily go to nest. One wild bird and a tamed bird will also nest readily. Two tame ex-pet birds will breed also, providing they like each other.

Birds are similar to humans: they are selective about their mates. If at all possible, let your breeder birds choose their mates. Unfortunately, this is not always possible, so you may have to look for a new mate for your bird if *your* first choice didn't appeal to it.

It is also important to realize that birds are *not* like dogs: Never put two birds together to breed only once. Once a pair bond is established and the two birds go through the nesting process, they will usually breed for you every year if you care for them properly. A bird that is allowed to breed only once or twice will feel lonely and sexually frustrated if its mate is taken away.

When you purchase birds for breeding, you should put them in separate holding cages until you are ready to introduce them. This does not apply to bonded pairs you may purchase. *Do not* put a single bird into a breeding cage with a nest box or it will become possessive of the nest box and may react aggressively to another bird. Put the two birds side by side in cages so they can look at each other. When you can supervise them, let them out on top of their cages, or on a neutral perch, one large enough for two birds to stand on. If they seem to fall in love instantly, you will know, because one may preen the other, or regurgitate and try to feed the other one, or the two may engage in mutual vent preening. If this is the case, you are probably safe in putting them in a breeding cage with a nest box right away. You may have eggs soon!

Most newly introduced birds will bicker and fight, however. In this case you should have a large squirt bottle full of ice-cold water and a long stick on hand. Every time the fighting becomes one-sided or too intense, try squirting them with the water, and if that doesn't work, then pick up one bird at a time on the stick and return them to their separate cages.

Pick them up with a stick instead of your hand so they don't purposely fight for your attention, getting you to pick them up. It's best if you have the birds in a semi-secluded area for their initial "getting to know each other" sessions, because sometimes birds fight only when *you* come around. This is especially true of Amazon Parrots.

During the introduction process and after the birds are set up in a breeding cage, you must make a commitment to *neither* talk to *nor* play with the birds. This is not easy for many people, but if you converse with your bird or handle it during this time, you will confuse the bird.

When its around you, it's happy, but when it's around the other bird it can't form a strong pair bond, which is necessary for successful breeding. We call this "divorcing your bird," because one party (usually the bird owner) feels awful.

We have a female Scarlet Macaw that laid clear eggs for two years because she was "in love" with my husband, Omar. She looked forward to seeing him every day at feeding time. Finally, we had to bring her in the house and slowly introduce a new mate to her. She kept refusing the bird, preferring Omar. Omar started picking her up with a stick, and telling her "No!" when she would come over to him. Finally, after one month of feeling jilted, she became interested in the new male bird. They produce babies for us every year now! This is an extreme case, however. When buying wild birds, you don't have to worry about such bird-human relationship problems.

Birds That Show No Inclination To Breed

Sometimes, despite all the right procedures, a pair of birds will show no inclination to breed. We try to give these birds one year to change their minds. After one year, we may move their cage location or change the nest box. One successful breeder I know recommends putting such birds in a carrier and taking them in the car for awhile (weather permitting) to "shake them up." Once returned to the safety of their cage, many of her pairs have had a change of heart!

It is also important to realize that birds that are more difficult to breed, such as Red-bellied Parrots, Mini-macaws, Hawk-headed Parrots and Caiques do not benefit from the stress of changing things around, but seem to thrive on a constant, reliable environment. Birds that are very timid and shy should be provided with calm, consistent environments. Sometimes a visual divider helps a pair of birds that seem to need extra privacy.

If your pair does not interact and you've let them stay together for a year or more, you may wish to try looking for another male or female. Do not sell the old mate until you are sure the new combination will work. Some pairs may appear to dislike each other, but once separated they act differently. We had a pair of Green-winged Macaws that for a year and a half sat on opposite sides of the perch. We never saw them preen each other or act as a bonded pair. We decided to sell the female and replace her with another that might be more productive. Two days after we delivered the female to its new owner (who placed it in a cage, without quarantining the bird, with a mature male Green-winged Macaw) the bird laid an egg off of the perch! As soon as we heard this we knew we had made a grave mistake; our female had been getting ready to lay with her old mate, and was already cycling to lay an egg when we sold her. We wanted to buy back the hen. The man we sold the female to simply thought that our female really liked his male, and that was why she laid an egg so quickly. We explained to him that the bird had not bonded with his bird, but had simply followed through on a physiological process that started at her original home, and previous mate. We were afraid that this fellow would not sell us the bird back, but eventually he did, and after we put the pair together again, nothing happened right away, but due to the stress of the entire situation, that is not uncommon.

CHAPTER 5

FEEDING YOUR BIRDS

Feeding your breeders the proper diet is one of the most overlooked aspects of bird breeding.

Birds that are not fed adequate diets can easily become egg bound, lay clear eggs, soft-shelled eggs or no eggs at all. Birds without proper diets also are more apt to neglect or eat their babies. DIET is important! If you cheat on your birds' nutritional needs you cheat yourself of all the babies you could have raised.

We feed our birds with three dishes, one each for seeds, fruits and vegetables, and water.

Important Note: Always feed your most valuable birds first, and feed your birds in the same order every day. This will help prevent the spread of a virus or bacteria to your entire collection. For example, if you have a mixed collection of birds, feed the Macaws before you feed your Ringnecks.

Seeds

The seed mix we use is a large hookbill mix that contains a variety of seeds: safflower, sunflower, buckwheat, pumpkin, oat groats, peanuts, popcorn, and Wayne® brand Puppy-O's.

It is important to use a seed mixture that includes a variety of seeds.

A lot of people are afraid to feed their birds sunflower seeds because they have heard that sunflower seeds are addictive and not nutritious.

Sunflower seeds are not addictive, nor do they have narcotic properties, but if your bird eats only sunflower seeds, it will ingest too much fat. Sunflower seeds can be good for birds that need to gain weight or baby birds learning to eat seeds for the first time. Many people feed a safflower-based hookbill mix to reduce the fat level of their birds' diets but safflower seeds are as fattening as sunflower seeds! That is why it is important for your birds to eat a variety of seeds, in addition to their fruits and vegetables.

If you do feed sunflower seeds, buy the gray-stripe variety. These are the highest quality. Black sunflower seeds are the lowest quality and should be avoided. Extra peanuts should be fed only as an occasional treat, because they are also high in fat.

In addition to their seed, we give all of our Macaw pairs raw, unsalted, large, fancy mixed nuts (walnuts, Brazil nuts, pecans, almonds, and filberts) in the shell. The Macaws eat these before anything else. Our Amazons and Greys enjoy these nuts, too, but we have to crack the larger ones for them. Sometimes we serve raw, unsalted pistachios or pine nuts, and the birds really seem to enjoy these.

Fruits and Vegetables

There should be no limit to the variety of fruits and vegetables you feed your breeders. The biggest mistake people make is to neglect to give their birds lots of raw, fresh fruits and vegetables. Even if your bird doesn't eat them at first, offer a variety every day. Eventually, the bird will nibble a piece of corn or maybe a carrot. Most birds will try new things when the female goes to nest, because the male has to feed her, which means he eats for two or more, and he is willing to be less selective. Many birds start eating fruits and vegetables when they have babies: This is a never-ending task, and soft foods, especially corn, are easily regurgitated to babies.

Remember: Birds that are served only seed will not learn to eat fruits and vegetables. Such birds may have shorter life spans than those that eat a variety of fresh food, as they would in the wild.

Every day we cut up 4-6 different kinds of fruits and vegetables in large pieces, and each pair of birds gets two pieces of each item. We change the "menu" every day, and the birds look forward to their fresh fruits and vegetables and never become bored with their food. We also cut up whole grain wheat bread in 1-2 inch pieces and each pair gets two pieces of that. Sometimes instead of wheat bread we serve rice cakes or unsalted crackers.

Any kind of fruit or vegetable is fine *except* avocado. Avocado has sometimes proven poisonous to birds and should never be served.

Do not serve the following fruit seeds to your birds: apple seeds, cherry pits, peach pits, apricot pits and mango seeds. They can be toxic.

The fruits and vegetables we serve most often are:

Corn on the cob *
Apple
Orange (seeds okay)
Banana
Mango
Papaya (seeds okay)
Grapes (seeds okay)
Pears
Peaches
Apricots
Grapefruit (seeds ok)
Strawberries
Coconut
Kiwi Fruit
Raspberries
Watermelon
Kumquats

Spinach leaves **
Brussels sprouts **
Carrots
Cucumbers
Alfalfa sprouts
Tomatoes
Zucchini
Dried red-hot chili peppers
Green bell peppers
Red bell peppers
Yellow bell peppers
Fresh chili peppers
Green Beans
Peas in the Pod
Broccoli
Yellow Squash
Summer Squash

* We try to feed this every day to nesting birds or birds with babies.

**These contain Oxalic acid, which prevents birds from absorbing calcium, so these items should not be fed more than twice a week. They still have many good nutrients.

We always serve large, raw, cut-up pieces of *fresh* produce. Frozen vegetables spoil in several hours and do not contain nearly as many nutrients. Cooked bean mixtures are nutritious, but also potentially dangerous because they spoil quickly.

An example of the way we serve fresh fruits and vegetables to our birds every day.

Vitamins

We use a powdered vitamin supplement such as Avia® or Nekton-S® on the seed every day. We sprinkle the vitamins lightly on the food as if we were salting it for seasoning on something we would eat. Many people do not believe that powdered vitamins work because the bird only consumes the inside of a seed, and not the shell. If you watch a bird eating seeds, however, you will notice that before the bird cracks open and eats the inside of a seed, it will stir up the seeds in its seed bowl, pick up a seed, roll it around in its mouth, and either drop it for another, or crack it open. All of this time, the bird is getting the powdered vitamin supplement on its beak and tongue. There are some vitamin supplements that are grainy in texture, and it is best to avoid this type of vitamin because it will settle in the bottom of the bird's food bowl. The majority of bird vitamin supplements, however, consist of very fine powder that will coat anything from seeds to fruits.

We also prefer to use the powdered vitamin supplements instead of vitamins in the water because they are more economical, and do not lose their effectiveness as quickly as a supplement that is added to the bird's water. Adding vitamins to the water can sometimes promote bacterial growth.

On the fruits and vegetables we sprinkle D-CA-FOS®, a calcium supplement manufactured by Fort Dodge Laboratories. African greys and birds that are laying should get D-CA-FOS® every day. We add the same amount of calcium to the fruits and vegetables as we do vitamins to the seed.

We also give large cuttlebones to pairs that are laying. We wedge the cuttlebone between the bars of the cage.

Pellets

Pelleted diets come in many different flavors, colors and textures. We occasionally mix a small amount of pellets in with the seed, but that is *it*. There are many potential problems with pelleted diets, such as spoilage, excess water intake, poor nutrition and boredom. There is also little data on the long-term effects of a parrot eating pelleted food; we know that birds have survived on seeds as long as we've known them, but they have only been eating pellets the last few years. Our main concern is that people will use pellets as a "crutch" so that they can justify not cutting up fruits and vegetables every day. Avian experts still do not know exactly what a parrot's specific dietary needs are, so until we do, it is our job to try and provide our birds with as many fresh foods as possible. Birds do not eat just one thing in their natural habitats, and they shouldn't be forced to do so in captivity.

Water

We serve each pair of our birds fresh tap water every day in a clean bowl. When using tap water, let the water run freely for about five minutes before filling your birds' water bowls. This should rid the water of any possible contaminants that may have settled in the pipes. It is not desirable to use water from a hose, but if you have no choice, let the water run 15 minutes to avoid contaminants in your water.

We have two sets of dishes for each pair of birds. When we feed, we swap dirty dishes for clean ones. We wash the dirty dishes in the dishwasher. If you do not have a dishwasher for your bird bowls and wash them by hand, be sure you use a different sponge for each collection of birds. For example, if you have a Conure room and a Macaw room, be sure to keep a different sponge (using different colored sponges helps) for each room. Store these sponges separately, too. Replace these sponges at least once a month. Segregating your feeding bowls and sponges helps prevent spreading disease from room to room.

You may serve your birds bottled water if you like, but be aware that not all bottled water is as pure as it claims to be. If you use a dispenser for your bottled water, clean the entire unit with bleach and water once a week to avoid contaminating your water with a dirty dispenser.

Some bird breeders boil their water before serving it to their birds or before using it to make hand-feeding formula.

Automatic watering systems may seem convenient, but we question their ability to stay free of bacteria.

CHAPTER 6

CAGES AND NEST BOXES

Aviary Breeding vs. Cage Breeding

Some of us have no choice in whether to decide to breed indoors or outdoors. If you have a lot of property, and it is zoned for livestock, you may wish to set up your breeder pairs outside in large aviaries, or flights. Climate is an important factor to consider, and if the temperature dips below freezing most of the winter where you live, then you must decide if you wish to build a convertible indoor/outdoor facility to allow birds access to an outdoor cage during the milder months, or if you wish to enclose all breeder cages entirely. If you live in a heavily populated area and wish to set up even one pair outside, check first with your local building department or zoning commission to see if there are any restrictions.

Many people, us included, live in the suburbs and must convert extra bedrooms, garages, and storage sheds into bird breeding areas. In this case, you fall into the category of "cage breeding," which means that your birds will be set up in large cages, but will not have room for flying. Many birds do very well in cages, but some become overweight from lack of exercise.

It is important to have two different levels of perches so that the birds can climb from one area to another and not be confined to just one perch. Some birds that are set up in large aviaries become overwhelmed by the large area of the cage, and some don't even use the entire cage, prefering to sit in one spot all the time. Previously owned pet birds are usually accustomed to small to medium size cages and may feel insecure in a large flight cage. African Greys breed poorly in large, long, flights, instead prefering cozy, smaller cages.

Cage Breeding

We chose to cage breed our exotic birds because we live in the suburbs and do not have enough space to put large flight cages in our yard. We would also be quickly reported to city officials for noise disturbances from our neighbors! Instead, we chose to convert a garage and family room into bird-breeding facilities; when these buildings were properly insulated for noise, our neighbors were not disturbed by our birds, even though they know we have birds. It is important to have a good relationship with your neighbors—tell them to let you know first (rather than the city) if they are bothered by noise. Some clever breeders have been known to give their neighbors a cute little Cockatiel or Lovebird to introduce them to the hobby!

If you have the room and the right climate for breeding birds outdoors, it is of course much more desirable for the birds to be outdoors. More important, however, is to boost the numbers of domestically raised psittacines.

Macaws, Cockatoos and Large Parrots

For Macaws and Cockatoos, we use a wrought-iron cage that measures 4′x 5′x 6′ (see photo). Bar spacing is usually about 3/4 to 1-inch wide. We have these cages custom made in panels, so they can be easily taken apart for cleaning, and they can fit through the doorways of our breeding rooms. The cages are powder-coated with lead-free paint. The front panel has one large door and a small feeding door that we attached holders to accomodate number 2-size ceramic bowls. The food bowl holders are designed so the birds cannot pick up and throw their bowls. The cage is held together by numerous bolts, which are installed with the head of the bolt toward the inside of the cage, the nut is on the outside so the birds cannot take them apart. We also have a tray made for the bottom to facilitate cleaning.

This breeding cage is very functional for cage breeding Macaws. The wrought-iron cage panels are bolted together, and the cage can be disassembled for cleaning .

Amazons, African Greys and Medium-Size Parrots

For Amazons, African Greys and other medium-size birds, we use a cage that measures 30″ deep x 36″ wide; bar spacing is usually about 3/4-1 inch. We use powder-coated wrought-iron, but these cages are not collapsible.

To save space, we usually order a double cage on a stand, like the one pictured below. If the birds are too close to the ground, they usually will not feel secure enough to nest.

Medium-size birds can be stacked on top of each other as long as they are not too close to the ground, or they may feel too insecure to nest.

Caiques, Senegals and Smaller Parrots

Even smaller birds such as Caiques, Senegals and Meyers parrots we use a 2′x 2′x 3′ wrought-iron cage. Bar spacing on these cages is about 1/2 inch.

If you do not have the money to order a custom-made cage, build your own or buy used parrot cages and modify them to your needs. (Used cages should be thoroughly cleaned and disinfected, and all new perches installed. If you need to touch up the paint, use a spray paint that is safe for baby furniture, and lead free.)

Nest Box Size

We basically use two types of nest boxes: the rectangular box-type and the L-shape, or "boot" box.

For large birds such as our Macaws and Cockatoos, we use a rectangular box that measures approximately 2′x 2′x 4′ (see photo). The entrance is on one end and the inspection door on the other. We make these boxes out of 3/4-inch plywood and replace them once a year or whenever they need it. You can reinforce the bottom or sides with wire, but since all of our birds are indoors, we don't worry about birds escaping.

We prefer to use the same type of nest box for our Macaws and large Cockatoos. It is also easier to check on the babies or candle eggs with this nestbox.

For medium-size birds, like Greys, we use the L-shape or "boot" box. The measurements are 24″ tall x 12″ wide x 24″ long. The entrance is at the top of the "L" and the inspection door is at the bottom, or lower half of the box. (See photo). A platform at the base of the ladder encourages nesting in the base. You must use only *large* square wire when building your ladder. Birds can catch their toes, toenails or feet in small-gauge wire. Always cut off any open-type leg bands before putting your birds in a breeding cage; these bands can easily catch on the wire. A bird trapped by a leg band may break or chew off its leg.

For smaller birds, we use a small version of the L-shape that is 14″ tall x 7″ wide x 14″ long. All of the other details are the same as the larger box.

The only hardware you need to purchase (aside from the wood) are the hinges, the hasp and screws or nails.

Medium-size birds, especially African Greys, do well in a L-shape, or "Boot Box."

Lining The Nest Box

We line our nest boxes with pine shavings. Do not use dirt, peat moss, cob bedding, mulch or other materials because they will encourage bacteria to grow, which in turn will negatively affect the eggs, babies and the parents. One time, we used sterilized dirt in our nest box, and we lost three clutches (a total of 9 birds) of Scarlet Macaws due to bacterial contamination.

Always put enough shavings in the box. We fill each box with about 5-inches of shavings. The parents will kick out, push aside and shred whatever they don't use. If you put too few shavings in a box, babies may develop a condition called Spraddle Leg.

Important Note: Do not rearrange the shavings when you inspect the nest box. This may upset the mother and cause her to neglect her eggs or chicks. Also, do not add any shavings if your hen is in the process or has just finished laying; you may upset the pair and cause them to destroy or abandon their eggs. Never put food inside the nest to try to entice the birds into the box — this teaches them to eat in the nest, which could become a bad habit and possibly lead to eating their eggs or babies.

Changing and Cleaning the Nest Box

We change the pine shavings in the box after we pull each clutch of babies. If it is your pair's first clutch of the season, you can quickly scoop out the dirty areas and replace them with fresh shavings without taking the box down. If this is your pair's second or third clutch, take the box down, vacuum it, scrape any dried feces and spray it with a disinfectant such as Nolvasan®, Kennelsol™, Enviro-Clean or bleach and water: one-half cup bleach to one gallon water. Be sure to have a heavy piece of wire and some bird-proof clips handy so you can secure the hole while you are cleaning or changing the nest box. The box will dry quickly. Refill it with fresh shavings and re-mount as soon as it is dry. If the box is extremely dirty, smells bad or is badly chewed, replace it. These boxes are fairly inexpensive to buy, or you can make them yourself.

Important Note: The nest box should always be mounted on the back of the cage, up high, and there should be a perch positioned in front of the entrance. This perch should run parallel to the box and be placed slightly below the hole. If it is positioned too high, the birds may not enter the box. The perch in front of the nest should be the highest perch in the cage.

Cage Location

Cage location is extremely important. Many birds do not breed well in busy areas such as family rooms, occupied bedrooms or hallways. Your chances for success are greatly increased if you provide your birds with a quiet area that is disrupted only once or twice a day, when you feed them. This is especially true for shy birds, such as African Greys. The more privacy they have, the happier they are. Wild birds feel especially vulnerable in a busy area and can be in constant fear if there is a lot of traffic in their breeding area. Birds that were once pets may feel confused and not want to breed if they are set up to breed in a living room or family room with lots of people around.

Empty bedrooms, well-insulated and ventilated garages and protected patio areas can be converted into prime breeding areas. If you don't have a spare room, even setting up a divider and making an area more secluded is an improvement. You could also make a bird shed outside, if the climate is warm enough year 'round. Excess heat can be dangerous to birds, however, so adequate ventilation is important.

An unused patio area can be an ideal location for one or more pairs or birds.

If you convert a room into a bird room, be sure to either install tile flooring or securely tape down a heavy tarp to help keep the floor or carpet clean.

All of our breeder pairs are set up indoors because they would be too noisy outside! We installed 4-foot Vita-lites® and fluorescent lights every four feet in each room. (You can buy 4-foot fixtures that hold two fluorescent bulbs at most large hardware stores that require no assembly and are ready to plug in as soon as you install the bulbs.) Vita-lites® are said to be the closest light source to natural sunlight and may help birds convert vitamin D for use in their bodies, which they need for egg production and overall health. We also know of people who breed birds indoors and use plain fluorescent bulbs. They have not noticed any problems with their birds. We put our lights on timers so they go on around 7:30 in the morning and off at 6:30 at night. We replace our Vita-lites® about every six months. The only time we change the hours is during daylight savings time. We do not control photoperiods (the amount of time we leave the lights on) to increase production. We provide lighting only because our birds do not get direct sunlight indoors. Be sure your lights are not accessible to the birds, or they will break them.

The advantages of cage breeding are:

- Bird breeders that live in the suburbs can breed parrots indoors.
- Because the birds are in an enclosed building, escapees cannot get farther than the building itself.
- Indoor bird rooms can be insulated to cut down on noise.
- Breeders can house more birds in a smaller area.

The disadvantages to cage breeding are:

- Most larger psittacines cannot fly in cages.
- Indoor birds require additional light supplementation.
- Some birds tend to become overweight with less room to exercise.
- Some breeders become greedy and breed too many birds in inadequate cages.
- Indoor birds cannot bathe adequately and must be periodically bathed or misted.

Important Note: It is possible to set up aviary-size flights indoors. There will be less room for more birds, but some bird breeders do not want a huge collection.

Aviary Breeding

If you wish to set up your breeder pairs in large flight cages, you may do so indoors or outdoors. Most bird breeders who breed their birds in aviaries do so in outdoor flights. In warmer climates, such as Arizona, California and Florida, the birds can be housed outside all year 'round.

It is best to introduce a new purchase (after its quarantine period) into an outdoor aviary in late spring or early summer. This way the bird has time to acclimate to the outdoors before it starts getting cold. If you place a bird that is not used to being outdoors in winter or early spring, you are increasing the chance that the bird will become ill.

Many states are warm and mild in the spring and summer, yet very cold in the winter; one breeder friend of ours lives in Oregon and has built a barn for his birds. The cages and nest boxes are inside the barn, and each cage has an outdoor extension with a trap door to the barn. In the cold winter months, this breeder shuts off the outside sections of the flight cages and uses fluorescent lights inside the barn.

As soon as the weather warms up in the spring, he opens the trapdoor to all of the birds' cages to allow them access to the outside section of their flight.

Another consideration you must take into account when designing your outdoor flights is security: Are the cages sturdy? Can a bird easily chew its way out of the cage? Will you need a extra, fenced-in section in case a bird does escape? Are your nest boxes secure? Will you fence in the nest box section for added security?

Be sure you have concrete flooring in your aviaries: it is easier to clean, and it helps prevent mice and rat infestations. I always suggest to people who wish to build outdoor aviaries that it is a good idea to make a second, larger enclosure with chain link, or a similar aviary-type wire, that will house many individual flights. The advantage to securing your aviaries in a second enclosure is that you will never lose a bird that has managed to get out of its cage, and you will decrease the chance that predators (such as raccoons, owls, hawks and coyotes) will come to your flights and injure or kill your birds. Your start-up costs will be much greater with the added expense of a second, larger enclosure, but think of what a loss it would be if a predator killed one or more of your proven breeders.

Cage Material

Most people who build large flight cages use either chain-link or wire that is galvanized after welding. Many cage wire vendors buy and import wire from England. It is important to buy the highest quality of wire possible; the wire should be smooth and not covered with a lot of flakes, chips or particles.

Galvanized wire also has a zinc coating to protect the wire from rusting, and some birds may develop zinc poisoning from chewing on and injesting loose flakes, chips and particles of the cage itself. (That is why the quality of the wire is important to bird breeders.) It is important to wire brush any new cages to remove any loose chips or flakes. Remove any particles from floor of the cage so the birds cannot go down and pick at them. Always wash a new cage thoroughly with soap and water before placing any birds in it. If one of your bird's becomes ill shortly after putting it in a new wire cage, remove the bird from the cage, and be sure to have your avian veterinarian check the bird for zinc poisoning, or New Cage Disease.

You should also be aware that cage wire could contain trace amounts of lead. The amounts are sometimes so small that they are virtually non-measurable. It is recommended that you send a sample of your wire to an independent laboratory so that they can test scrapings of the cage for lead and zinc content. Some sources say you should rinse all new cages with an acid wash, for example: vinegar and water, but that will only remove impurities from the surface of the cage, not the pure metals that make up the wire itself. A breeder friend of ours soaks all new rolls of wire in a straight vinegar solution for one-half hour as a preventative measure.

When choosing cage and aviary wire, be sure that the wire is strong enough to withstand the bird you intend to put in that cage. Most wire companies are glad to offer suggestions as to what gauge of wire you will need.

Nest Boxes in the Aviary

In cage breeding, the nest box is installed on the outside of the breeding cage; in aviary breeding the nest box is usually mounted inside the flight, against one wall, with access to the nest through an inspection door. The nest boxes would be the same size as those I described in the cage-breeding section. The box should be hung in the highest section of the cage, and the highest perch should be installed just below the opening to the nest box.

Shelter

You must provide some sort of partial roof or shelter from the elements over the flight cages. This sheltered area must also cover the nest box area. This will shelter the birds from rain and direct sunlight.

Cage Size

Small parrots, such as Caiques, Senegals, Meyers, and Spectacled Amazons are usually housed in cages 3 feet wide, 4 feet long and 6 feet tall or larger.

Medium-sized parrots, such as Amazons, Bare-eyed Cockatoos, and Ringneck Parakeets, are usually housed in cages 4 feet wide, 6 feet long and 6 feet tall. Please note: African Grey Parrots are awkward fliers and prefer smaller cages.

Large parrots such as Moluccan Cockatoos and Macaws seem happiest in a cage at least 4 feet wide, 8 feet long and 8 feet tall.

- Birds can free fly in an aviary, reducing the possibility of becoming overweight or out of shape.
- Birds can bathe in a natural rain shower.
- Birds have access to natural sunlight.

The Disadvantages of Aviary Breeding:

- Unless there is a secondary enclosure, escaped birds can be lost, and all birds are at a higher risk to outside predators.
- Insects and rodents may infest the aviary.
- Extremely hot weather may cause heat exhaustion, or overheating of eggs in the nest.
- The start-up cost is more expensive than for cage breeding.
- Noise and space limitations make this method unfeasible in a suburban area.

Outdoor aviaries must provide the birds with a roof for shelter.

CHAPTER 7

EGGS

Indications of Egg Laying

It is very common to see a pair of birds entering and spending time in their nest, but not producing eggs. It is also common for birds to copulate, or mate, frequently but fail to produce eggs.

Some birds copulate only before they nest; every pair is different. This is why it is important for you to keep records on your birds. A daily journal is good, or a large wall calendar works, too. When you first start breeding, record anything that might be pertinent to your pair, such as "birds started eating more vegetables; female spending time in nest; or pair more aggressive, etc."

Here are several indications that your pair may be getting ready to lay:

- The female stays in the nest a lot, or all day.
- The female passes large stools, sometimes blackish-brown.
- The pair is more aggressive toward you (or each other) at feeding time.
- The male feeds the female more often than he did before.
- The female appears swollen in her vent or lower abdomen.

Important Note: One or more of these signs can indicate a sick bird. That's why a thorough veterinary exam is *essential* before you set up your birds.

Frequency With Which Eggs are Laid

As soon as your hen is cycling, she will start laying eggs in the nest box. The size of the bird usually determines how often it will lay its eggs. Small female parrots usually lay an egg every other day, in the early morning hours. Larger species such as African Grey Parrots, Amazon Parrots, Macaws and Cockatoos will lay their eggs up to three days apart. For example, a Macaw may lay its first egg on Monday, the second on Thursday, and the third on Sunday. Sometimes an inexperienced hen will lay only one egg, or will lay eggs anywhere from 2-7 days apart. Don't worry if your birds don't start incubating (sitting on) the eggs right away; a pair may wait until the hen is finished laying. Then all the eggs will hatch around the same time.

If your hen repeatedly lays her eggs outside of the nest box, either on the floor of the cage, in a food bowl, or off the perch, you must make some changes to the existing nest box (or try another design) to encourage the hen to lay her eggs inside. If you find a broken egg on the floor of the cage, pad the bottom of the cage with a thick layer of pine shavings or towels in case the hen should drop any more eggs she may lay. The hen may also have laid the egg inside the nest box and then either she or the male threw it out!

Normal Clutch Size

The expected number of eggs in a clutch also varies in relation to the size of the bird. Smaller parrots such as Caiques and Senegals usually lay at least three or more eggs. Amazons, African Greys and other medium-sized birds will lay between two to four eggs. Large Macaws and Cockatoos usually have one to three eggs in a clutch.

Clutch size may vary according to individual pairs, also. We have a proven female Blue and Gold Macaw that lays anywhere between three to five eggs, and usually almost all of them are fertile. We also have a proven African Grey female that only lays two eggs, which are usually both fertile.

Important Note: If you have to remove eggs from a female for incubation as soon as they are laid, she will probably lay one or more eggs than if the eggs had been left with her in the nest.

Infertile Eggs

If the pair has been copulating and making good contact, the sperm can enter the female, and most or all of the eggs should be fertile. However, there are lots of reasons for infertile eggs. One of the following problems is usually the cause of infertility:

1) You have two females instead of a male and a female.
2) One or both of the birds is too fat.
3) One or both birds has a bacterial infection.
4) The male has low sperm count.
5) At least one of the birds is too young or too old to reproduce.
6) The birds' reproductive cycles are not occurring at the same time.

Candling The Eggs

Once the eggs are laid and the hen has started incubating, you can wait 5-7 days to check the eggs to see if they are fertile. This is commonly referred to as "candling the eggs."

The easiest way to candle eggs is with a small, flexible flashlight. There are many types of flexible optic lights available, but we use an inexpensive auto map light, which you can buy in any auto-supply store.

To candle the eggs, you must cover the entrance to the nest box while the pair is outside. This can be quite difficult because once the pair is on eggs they will not get off the nest often or readily. Sometimes you must try to coax the pair out of their nest box. Some birds, however, will readily run out of the nest, while others will remain stubbornly inside, making it extremely difficult for any type of inspection. In the latter case, try placing a magazine or piece of wood between the birds and the eggs. (Watch your fingers!) Sometimes this is a two-person procedure; I will get a pet bird from the house and hold it in front of the breeder cage, while my husband quickly opens the inspection door and candles the eggs. Usually both the male and female will run out to attack their intruder, but as quickly as they run out to charge, they run back to their nest.

Of all of our pairs, the Cockatoos are the easiest to deal with; they usually back out of our way, while hissing and swaying to try to scare us. All of our large Cockatoo pairs are set up with horizontal Macaw nest boxes, so they have room to move away from us while remaining in the nest box during candling. When we used the T-shaped nesting box with two entrance holes, we had lots of problems with the birds trampling or puncturing the eggs during candling, because there was no place for them to go, and they would panic and break their eggs.

To candle an egg, hold the light up to the wide end of the egg; you will see the air space and the contents. If there are any red veins present, the egg is fertile. If the egg looks clear inside, the egg is probably infertile. Don't mistake streaks on the eggshell for veins. Do not discard any egg until you are absolutely sure it is infertile. (Check the egg 2-3 more times before pulling it, if your pair will allow it.) Continue candling fertile eggs every 4-5 days to check their progress. As the embryo continues to develop, the interior of the egg becomes darker and more dense. If your light is strong, you can still see faint veins encircling the egg.

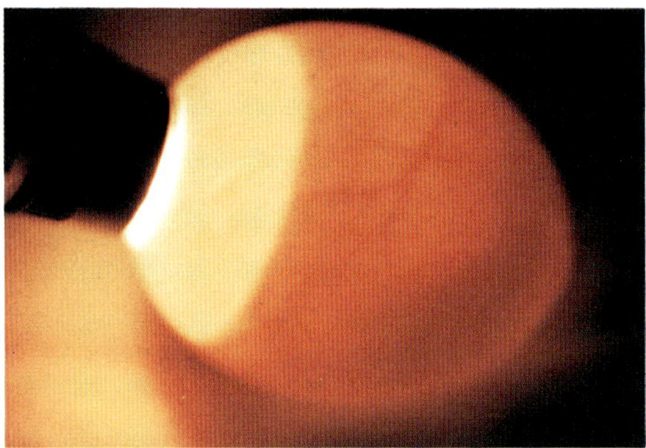

Candling a fertile egg.

Incubating The Eggs Yourself

Ideally, every breeder should let the birds incubate and hatch their own eggs. However, not all birds take care of their eggs, and not all breeders are that trusting. If you are going to breed exotic birds, you should invest in a good quality incubator. If you can't afford a deluxe version, then start with a small, simple model until you have enough money to buy a good unit. We currently use the Humidaire Model-A, and the Grumbach®. Many breeders start out with a small dome-type incubator such as the Lyon Electric Turn-X.

Different incubators use different turning methods. The Turn-X turner rolls the eggs back and forth every hour. The Humidaire holds the eggs in a tray that rotates 45 degrees back and forth every hour. The Grumbach® slowly rolls the eggs a full 180 degrees as often as you program it to do so. Parrot eggs do best when they are turned 180 degrees about five times a day, alternating the direction. Never turn the eggs in the same direction: It weakens the chick.

If your incubator rocks the eggs back and forth on a tray like the Humidaire, you should turn each egg by hand, 180 degrees one to three times a day. So you do not forget which way to turn an egg, write carefully with a pencil on the egg: on one side, write a letter "A" and draw an arrow pointing to the right; on the other side, write "B" and draw an arrow to the left. That way, whoever turns the egg knows which direction to move them.

Always make sure your hands are clean when handling eggs! Bacteria and viruses can enter through the porous eggshell and harm the chicks!

Temperature

The ideal temperature for incubating parrot eggs is between 98½ to 99¼ degrees Fahrenheit. Invest in a large, good-quality thermometer (the kind that reacts quickly to temperature change, either up or down) and test the temperature in your incubator. The hatching tray is usually a bit cooler than the top of the incubator, but that is normal. If you have a plastic-type incubator, the temperature will fluctuate with changes in the room temperature, so try to pick a room that stays the same temperature.

If you leave the eggs in the nest with the parents for 10 days, your chances for 100% hatch rate are much greater than if you pull them when they are laid, providing they are fertile, of course.

Earthquakes, Storms or Other Disturbances

If your pair is scared off the nest during an earthquake, a storm or another disturbance, and they do not return to the nest in 15 minutes, you will have to pull the eggs and incubate them. If the ambient temperature is warm, you may leave the eggs in the nest box for up to 30 minutes. It is possible to pull and incubate the eggs until the birds have calmed down and return them to the nest, but we have only done this with pairs that are normally excellent parents.

Chipped or Cracked Eggs

If you find a chipped or cracked egg, repair the egg with clear nail polish to seal the cracks. Put nail polish only on the damaged areas. Be sure the polish is dry before you put the egg back in the nest box.

Although you may wish to attempt to repair a severely damaged egg and incubate it, such an egg usually does not develop correctly, and bacteria may grow inside the egg and damage or kill the chick.

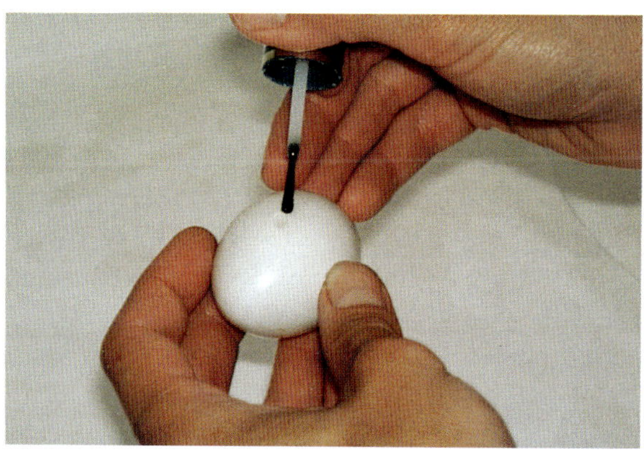

Repairing a chipped egg with clear nail polish.

Pipping

When an egg is almost ready to hatch, the airspace in the wide end of the egg shifts and tilts to a 45-degree angle. The border between the embryo and the airspace looks softer and not as defined. This is called *internal pipping*: The chick breaks the membrane with its egg tooth (a horn-like bump on its beak) and starts breathing the air inside the airspace.

When the chick runs out of oxygen, it breathes carbon dioxide, which causes its neck muscles to jerk in spasms. The head and beak then push up and break the hard shell of the egg. This is called *external pipping* and usually occurs within 24 hours after internal pipping. The actual hatching process may take one hour or 12 hours. The chick must push with its legs, neck and beak against the perforated end of the egg until the eggshell pops open, and the chick can climb out.

Unfortunately, many things can go wrong, especially if the parents did not incubate the eggs.

Sometimes a pipped egg is mistaken for a chipped egg; check your dates.

Birds That Pip Internally but Do Not Pip the Shell

This common occurance is usually caused by low humidity during incubation. If you notice the chick has not pipped externally 24 hours after internal pipping, you should make a small hole in the air space to see what's going on. Candle the egg to locate the chick. Use a plastic toothpick or a small knife (or anything that can be sterilized before using) to make a 1/4-inch size hole. Look inside. If the membrane looks white and stiff like paper, the baby is too dry and will need help hatching.

Assisting a chick that internally pipped but could not break the shell.

Enlarge the hole and dip a cotton swab or paintbrush in warm, distilled water and lightly dab it on the membrane. Try to locate the chick's beak and see if it has torn the membrane around the beak area. If it has, try pushing back the membrane around the chick's beak and nostrils. If the baby peeps loudly, it is usually the sign of a strong chick.

Assisting a Hatch

If the chick appears slow moving, and slowly opens and closes its beak, it is probably weak from struggling against the hard membrane. Put the egg back in the incubator and moisten the membrane every hour. When assisting a hatch, avoid all blood veins when peeling back the membrane. Breaking a blood vein could cause a chick to bleed to death. Avoid the chick's nostrils, and keep trying to push back the membrane to expose the chick's beak and head. If there is any bleeding, STOP, and try again in a few hours; the chick is not ready to hatch. Continuing to assist such a hatch may kill the chick. If there is no bleeding, you may continue to peel back the shell, starting with the area over the airspace.

Anytime you assist a hatch, if there is any bleeding you must stop immediately.

It is important to realize that there are no set rules to follow when it comes to hatching eggs. Each situation is different, and the guidelines in this book may not apply to your particular needs.

Once the baby's head and shoulders are exposed, turn the egg sideways so the baby can crawl out, with the help of gravity. If the baby is very weak, you may need to gently lift its head out. If you do this, look inside at the chick's belly to make sure it has absorbed all of the egg yolk. If it has not, you will see a greenish-yellow lima-bean or pea-like ball stuck to the chick's belly, between its legs. If you see this, you must put the chick's head back and cap the egg with another eggshell—one that has been thoroughly cleaned and rinsed with a nontoxic disinfectant. Try to cap the egg loosely, so that air can still get in, but the baby must remain confined inside the egg. If you don't want to use an eggshell, you can use plastic wrap that has been disinfected and poked with several holes. Be sure to poke the holes before placing it loosely over the end of the egg. Leave the egg alone for 6-8 hours and then check it again (moisten if necessary). If the yolk is still visible, then you must wait. If the yolk has been absorbed when you check again, you may leave the egg uncapped and continue assisting in the hatch.

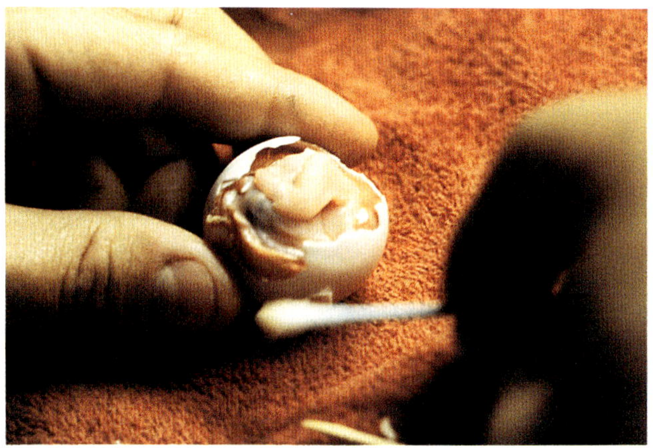

Before continuing to assist the chick, make sure it has fully absorbed its egg yolk.

The baby may still be connected to the egg by a thin cord to its umbilicus. Allow the chick to naturally break this. There will also be a small mass of green to clear waste products inside the egg. The first time the baby passes a stool, it is bright green and very gooey looking. This should not be confused with the waste products in the shell. But you may see both in the shell after a baby hatches.

If the baby is still connected to the shell by the umbilicus, leave the egg next to the baby in it's basket. It will naturally break this cord on its own.

CHAPTER 8

CARE AND FEEDING OF THE BABIES

Babies In The Nest

If you decide to leave the babies with the parents, it is your responsibility to be sure that they do a good job. You must check on the babies in the nest once or twice a day. The best time to check is about an hour after you feed the adults, because the parents usually feed the babies right after they eat.

The baby's crop should be partially or completely full when you check. The baby should be in an upright position and its skin should be light pink. If the baby is hungry, it will cry to be fed. If it is full, it will be sleepy and quiet. If the baby is cold or cool to the touch, it should be pulled for hand-rearing. If the baby has not been fed for 24 hours and appears weak, pull it.

Assuming the parents have fed the babies, the best time to pull babies for hand-rearing is when their eyes start slitting open, which on a large bird, such as a Macaw, happens around 12-14 days after hatching. African Greys' eyes open around 10-12 days, and smaller parrots open their eyes sooner. If you have a clutch of three babies, pull them when the middle baby's eyes are opening. It will not make a difference if a baby's eyes have been open several days, or even a week, before it is pulled.

Environment for the Day-1 Baby

When you pull a baby, immediately put it in a thermostatically controlled environment, such as an incubator or brooder. If you have only one incubator and it has eggs in it, you may keep the babies in the hatching tray for about a week. (The temperature in the hatching tray is usually a degree or two lower than the top and middle of a larger incubator, but even if it isn't, 98 degrees Fahrenheit is usually fine for most new babies).

Important Note: Some birds require less heat than others, so it is important to look for signs of overheating, such as panting.

If you have a separate brooder, set the thermostat at about 98 degrees Fahrenheit. After the chick is seven to ten days old, gradually lower the temperature, approximately two to four degrees each week. If there is more than one baby in the clutch, the babies generate warmth, and huddle together. If the babies are too cold, they will feel cool to the touch and will shiver; if the babies are too warm, they will separate and spread out. Continue gradually adjusting the temperature of your brooder unit until you find one that suits the group of babies.

Babies fed by their parents usually weigh more, cry less often, and maintain their body temperatures better than babies that are hand-reared from hatching. If your babies cry all the time, they may be too warm. Some babies cry when they are hungry and even after they are fed. This crying is different from when they are too warm, which is a more distressful call.

Making Your Own Brooder

If you have to pull a newly hatched chick, and you don't have an incubator or brooder, you can make one using a dishpan, heating pad or hot-water bottle and towels.

First, put a towel in the bottom of the dishpan. Then place the heating pad or hot-water bottle on one side of the pan. Cover it with another towel. Be sure that any time you use towels around baby birds to use velour-type towels, or tightly woven terrycloth, so that the babies cannot get tiny threads wrapped around their toes or toenails. Then make a little "nest" with a clean T-shirt (or another towel). Place about four tissues inside the nest and scrunch the bottom, so it's not slippery.

Put the baby inside the nest and cover the dishpan with a heavy towel, leaving about a 1-inch gap on one side. Make sure you use only the *Low* setting on the heating pad. Anything else is too hot. Fill the hot water bottle with water that is approximately 100-degrees Farenheit; always check the temperature by placing a thermometer made for human use inside the "nest." The temperature may fluctuate throughout the day, so you must check thermometer often; refill water bottles before they cool. Everytime you take the baby out to feed it, change the tissue paper.

To add humidity to your homemade brooder, soak a clean, new sponge in a shallow dish of water, or fill a small baby-food jar with water and poke several holes in the lid. Do not use an open container of water; A wandering baby may fall in it and become chilled or drown.

When the babies get older, you may wish to use something more absorbent than tissues for the droppings. We have used pine shavings for many years, but you need to be sure the babies aren't eating them; if they do, their crops can become impacted, and it may be necessary for a vet to draw out the shavings or remove them with forceps. To avoid this problem, we simply tear paper towels or napkins into 1-inch strips and cover up the shavings with the paper strips. You can also try using pressed and cleaned pine shavings; they are lighter, softer and more easily digested than the hard shavings, which can have sharp edges.

There is also a new bedding product, called Care-Fresh™, which is made from recycled materials and is biodegradable. It is also digestible. We now use this type of bedding with all of our babies, and it is especially convenient with larger Macaws that like to pick at their bedding, even when they are very young. We do not have to remove this bedding from their crops because it digests quickly and is eliminated in the stools.

Once the babies are older, you can put them on a grill so they don't have access to their feces or bedding materials.

Important Note: When using any shavings or litter material, place newspaper underneath to absorb excess fluid.

Traveling With Babies

Baby birds that are in good health travel very well and with little stress as long as you keep them warm and fed. Instead of a heating pad, you may use a hot-water bottle that you refill every few hours. If you are going to be traveling by car, you can purchase an AC adapter so you can plug your heating pad into the car's cigarette lighter and maintain the same temperature at all times. I used to take my babies to work with me and never had a problem.

Hand-feeding The Babies

Babies that are pulled from the nest can be fed as soon as their crops are empty. Babies that are hatched in an incubator or pulled right after hatching are still absorbing their egg yolk and should not be fed solid foods for about 24 hours, or until they stop passing feces. You may feed these babies a few drops of warm water, Pedialyte® or Lactated Ringers Solution every 2-3 hours. Babies will begin passing urine only (no dark green feces) when they have absorbed their egg yolk, as long as they are being fed fluids.

Important Note: If you are late or accidentally skip a feeding, it will not harm the baby. Parents are sometimes late, too.

Hand-feeding Technique

If you are right handed, hold a baby's head with your left hand, with your thumb on one side of the beak and forefinger on the other side of the beak. If you are left handed, hold the baby's head with your right hand and feed with your left.

Insert the syringe in the baby's mouth. Each time you feed, try to alternate; feed from the right side sometimes, other times from the left. This will help prevent the beak from curving, which is especially common with Macaws. If you alternate sides while feeding and still have problems with crooked beaks, your hand-feeding formula could be the culprit.

Some breeders try to hand-feed only from the bird's right side to prevent the food from going down the windpipe. The truth is, if the bird is hungry and showing a strong feeding response (bobbing), then you may feed from either side, because if the parent bird were feeding the baby, it would regurgitate into the baby's mouth. The baby instinctively closes off its windpipe, lessening the possibility of aspiration.

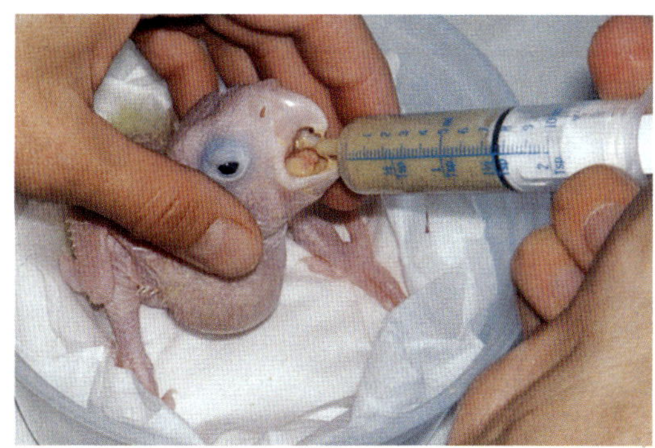

The proper hand-feeding technique.

Gavage Feeding

Gavage feeding, or tube feeding, is a technique that can be fatal if done improperly. If you are gavage feeding, then you must be sure you have the feeding tube in the esophagus before you begin the flow of formula, or you will aspirate the baby (get formula into its lungs).

Gavage feeding is sometimes necessary for birds that are ill and must be force-fed. Be sure you know what you are doing before you attempt to gavage feed. There are soft rubber tubes and curved metal tubes available. We have used both and prefer the curved metal tube with the ball tip. This type of gavage instrument can also puncture a crop if not used properly, so it is vital that you have your avian veterinarian demonstrate for you. Once you get the hang of it, it is really quite simple.

Some bird breeders gavage feed because it is a faster method to feed. We feel that it is more beneficial to syringe feed a bird because it does not take much more time, but gives the bird the opportunity to naturally bob for its food if it wishes to do so.

There are some exotic birds, such as Eclectus Parrots, that become extremely difficult to syringe feed. If the bird is too young to eat soft foods on its own and is losing weight, then it is beneficial to gavage feed birds such as these so that they can continue growing.

Amount to Feed

Probably one of the most misunderstood concepts in hand-rearing is how much to feed. There is no set amount, but you must think of the bird's crop as a measuring tool. The skin of the crop is very pliable and easily stretched, so you don't want to overfeed, because then the crop becomes pendulous, or stretched out. On the other hand, some people starve their birds, because they don't realize the baby's crop is empty. The folds of skin in the crop sometimes look like food. If you are not sure when your baby's crop is empty, add some spinach or other green vegetable baby food to each feeding. The green will make it much easier to tell if food remains in the crop. (Do not feed spinach as the primary source of food for more than one or two feedings, however, because it binds calcium.)

In the wild, parent birds never wait until the baby's crop is empty before feeding, they feed the baby, whether its crop is empty or not. If your baby has a *small amount* of food in its crop the next time you want to feed it, go ahead and feed it at that time. You may wish to feed slightly less formula so it will be ready for its next scheduled feeding. When a bird digests its food too slowly, it *may* have sour crop, or a bacterial infection.

Feeding Guidelines for Babies Fed From Day-1

	Small Parrots	Medium Parrots	Large Parrots
Day 1-3	Every 1-2 hrs. 1/8- 3/4 cc's	Every 1-1/2 - 2 hrs. 1/2 - 2 cc's	Every 2 hrs. 1/2 - 3 cc's
Day 4-13	Every 2-4 hrs. 1 - 5 cc's	Every 2-1/2 - 3 hrs. 3 - 10 cc's	Every 3-4 hrs. 4 - 15 cc's
Day 14-21	Every 3-4 hrs. 4 - 10 cc's	Every 4 hrs. 10 - 25 cc's	Every 4 hrs. 15-25 cc's
Day 22-35	Every 4-6 hrs. 8 - 20 cc's	Every 5 hrs. 20 - 50 cc's	Every 6 hrs. 25-60 cc's
Day 36 thru Weaning	Every 8 hrs. Approx. 20 cc's and decreasing at weaning.	Every 8-12 hrs. Up to 70 cc's and decreasing at weaning.	Every 8-12 hrs. Up to 120 cc's and decreasing at weaning.

These guidelines are approximations for different types of birds. Every bird is different!

When you pull a parent-fed baby, you can still follow the table, but you must look at the next age group, older than the bird's actual age. For example, if you have a 2-week-old, parent-fed African Grey baby, follow the feeding guidelines for a 3-week-old baby. Parent-fed babies usually grow faster, and their crops can hold more food.

Don't make the mistake of pulling a parent-fed baby and not feeding it enough. Eventually the baby will become dehydrated and ill, and then it will become difficult to keep the baby well.

If the crop seems stretched out, it won't hurt to put a push-up "bra" on the baby to ensure proper digestion. See Chapter 11-Problems With Babies for more information about what to do for an overstretched crop.

Choosing a Formula-Ready to Use Formulas

We no longer mix our own handfeeding formula because we have tried the ready to use formulas and found them to be comparable in nutrition and much less time consuming.

We have tried many brands of ready to use hand-feeding formulas on different types of birds. We prefer a formula that is available in different fat content levels because different birds have different nutritional needs. Macaws and Amazons and Day-1 through two-week old incubator raised babies do well on a diet higher in fat (approximately twelve percent). Cockatoos, especially large species, gain weight too quickly on a twelve percent fat formula, (resulting in the possibly fatal fatty liver syndrome) and require a lower fat diet (approximately eight percent). I hesitate to mention any specific brand names, because we may at any given time decide to try another brand, or sometimes the manufacturers change their formula or its packaging, and it may influence our decision on which formula we use. We have cultured two of the most popular brands of hand-feeding formulas to test them for bacteria, and the results were very good. There was no harmful bacteria present in the batches that we tested.

With any diet you may wish to experiment as we have done until you get the proper weight gains you desire for your particular birds.

Commercially prepared hand-feeding formulas are a big industry. Unfortunately, many are not complete diets. How do you know if a formula is good? You don't really. No one fully understand the nutritional needs of most adult parrots, much less their babies. One option is to make your own formula, but if that is not possible, do some homework and ask breeders what hand-feeding formula they use. Be sure to ask people who raise the same type of birds you do! Cockatiels and Lovebirds may do great on one brand of food, but a Cockatoo or Macaw may not grow properly on the same formula. African Greys between the ages of 10 to 20 days do not do well on diets with too much protein, and may have pink, blood-tinged urine if fed such a diet.

Hand-feeding Recipe

Some bird breeders still prefer to mix their own handfeeding formula at home. This is the best homemade recipe we have used:

3 cups of water
25-30 Baked Zu-Preem® Biscuits (made by Hills-Science Diet)
1 jar Mixed Fruit and Yogurt baby food
* 1/4 tsp. Avia® vitamins
* 1/4 tsp. D-CA-Fos® (Calcium Supplement by Fort Dodge Laboratories)
** 1/8 cup Peanut Oil
** 1 heaping tbsp. Peanut butter

* Omit for babies under two weeks.
** Omit for baby Cockatoos and Eclectus (they require a lower fat diet).

First, bake the Zu-Preem® in a 350-degree oven for 20 minutes (or until lightly browned) to kill any harmful bacteria.

Next, blend all of the above ingredients to a grainy texture. Add water if the mixture appears too thick. Add more Zu-Preem® if the formula is too thin. Use fresh food at each feeding; do not re-heat any food left over from a previous feeding. We have refrigerated this mixture for up to 24 hours without any problems with bacteria growing. If you are feeding a lot of babies it may be more convenient to mix a fresh blenderful at each feeding.

Some avian veterinarians that are also bird breeders may offer suggestions and recipes you may wish to try; it is important that you do not jump around from one recipe to another. Try to feed your babies all the way through on one formula, unless you think it is inadequate.

Use of Antifungals

You should never use an antifungal such as Nystatin, Nilstat® or Mycostatin® as a routine procedure. Antifungal medications should only be used to treat true yeast and fungal infections. Your goal should be for your babies to digest their food on their own, not with the aid of an antifungal. Birds that are given these medications on a regular basis can actually become immune to them, which can make it very difficult to treat a subsequent infection.

The only time we use an antifungal medication is when a baby has been cultured and found to have *Candida* or another type of yeast infection that is sensitive to it.

Adding Additional Ingredients to Instant Formulas

We recommend against adding a lot of extra ingredients to an already "complete" commercially prepared food. You may disrupt the balance and original intent of the food.

Adding extra vitamins is dangerous, too, because you can easily "over-vitaminize" your babies. This results in stunting, poor feather quality, and stress lines on the bird's feathers, or other more serious health problems.

If you wish to add some human baby food to an instant formula for flavor, try to add less than 20 percent of the total volume.

Adding Antibiotics to Formula

We add antibiotics to the formula only when a baby is ill, and we culture the chick to find out which drug to use. Each antibiotic is different, but sometimes it is more effective to administer a small amount of baby food with the antibiotic one hour before a full feeding is scheduled. This way the bird's crop has a chance to quickly absorb the medicine.

Never add antibiotics to the babies' food as a preventative measure. This only lowers the bird's resistance to additional bacteria, and it allows the bacteria to build a resistance to that particular drug so the drug may not be effective later.

Weighing Babies

When you hand-feed one or more babies, weigh each one daily to check its progress. Be sure you weigh a chick when its crop is empty, and at the same time each day, to get an accurate pattern for growth. Until weaning, your babies should gain weight every day. If you feel the keel bone (the bone that runs down the center of the bird below the crop), you can get a general idea if a bird is thin. The bird's chest muscles should be firm on each side of the keel bone, and the bone should only protrude slightly. Weighing a baby, however, is the best way to catch a potential problem.

You don't have to buy the most expensive scale, but try to get the most accurate one available for your budget. We use a triple-beam balance scale, which is accurate to 1/10 of a gram. (One pound equals 454 grams.)

If you raise large parrots, make sure your scale goes up to 5 pounds or has a weight kit to weigh objects up to 5 pounds. Make your own weight chart or copy the one in the back of the book. If you need to take the bird to the veterinarian, be sure to take the weight chart with you.

Birds that gain slowly or not at all are almost always ill, and if they are not seen and treated by a veterinarian as soon as you notice a problem, they will usually die.

Introducing Babies to Soft Foods

Babies may be given soft foods and millet spray when they are feathered on 50 percent or more of their bodies. Start weaning them from the hand-feeding formula when they are fully feathered. When you serve fresh fruits and vegetables, cut them in *large* pieces, so the bird cannot swallow them whole. Do not feed frozen mixed vegetables or canned vegetables because they spoil very quickly. Here's a list of food to serve to weaning babies:

- Rice cakes
- Unsalted crackers
- Wheat bread (plain or toasted)
- Cheerios®
- Large (3-inch pieces or bigger) pieces of apple, orange, banana, zucchini, broccoli, carrots, cucumbers and corn-on-the-cob
- Millet spray
- Pelleted food

When the baby is feathered over 75 percent of its body, you may also put a dish of seed and a small dish of water in its cage. The chick will play with the seeds and get them dirty, but that's okay. Just be sure and change the seed frequently. You want the baby to get used to having seed around. It will soon start sampling different seeds. The first time the baby drinks, it will probably cough and sneeze. This is normal. Take away the water and reintroduce the bird to water in a few days.

By the time the baby is fully feathered, it will start refusing to hand-feed. As long as the chick's weight is good and it eats some soft foods on its own, begin weaning it. If the baby has been hand-feeding 3 times a day, begin feeding it 2 times a day. If the chick is hand-feeding only twice a day, eliminate the morning feeding and feed it only at night.

The best way to eliminate a feeding is to gradually reduce the amount of formula fed over a period of several days.

Birds That Refuse to Wean

Some birds will hand-feed as long as possible because they enjoy the ritual. On the other hand, another chick may not wean because it is sick. Take a chick that refuses to wean to your vet for a check up. The vet may culture a bird that will not wean. If it is sick, you must treat it before weaning it. If it's not sick, you can start gradual weaning.

If a healthy bird is hand-feeding 3 times a day, cut down to two feedings 12 hours apart. Add lots of goodies to the cage for the baby to nibble on!

If a bird is on two feedings a day, start reducing the morning feeding 5 cc's each day. For example: If a baby eats 35 cc's in the morning, give it 30 cc's the next day, 25 cc's the day after that, then 20, 15, 10, 5, 0!

Continue feeding the baby in the evening for about a week. Check the weight every day. A slight loss (10-15%) is okay, if the baby loses more than that, resume additional feedings. Then, follow the same procedure for the evening feeding.

After your baby has weaned, continue weighing it daily for about 3 weeks, but don't get out the syringe and hand-feed if your bird looks at you with sad eyes, begging. If you do, the chick will learn to make baby noises and beg all the time, just so you will hand-feed it.

Once your baby is weaned, you can weigh it once a week if you like, or just feel its keel bone.

CHAPTER 9

PROBLEMS OF BIRD BREEDING

Noise

Breeding exotic birds can be frustrating, exhausting and puzzling. One of the most common problems is noise. Many bird fanciers cannot set up some of their favorite larger breeds because of potential noise problems.

Check with the city you live in to see what types of restrictions there are regarding outside aviaries. Many cities also limit the number of animals you may own.

Vermin

Vermin and rodents cause a lot of problems, to both inside and outside aviaries. Rats and mice carry several kinds of harmful bacteria (*Salmonella* especially, which is highly contagious) and are very difficult to eliminate. Rodents will eat your birds' food, drink their water, defecate and roam in their cages and on top of and sometimes inside their nest boxes. Needless to say, this is bad for fertility and the health of your breeders.

You cannot use poison to kill rodents because sometimes they will carry it into your birds' cage or aviary. Mouse and rat traps set outside of a cage will help, but the best method to eradicate a rodent problem indoors is to tear out all of the drywall in your bird room, clean out the walls, and replace them with new material, making sure to seal any holes or gaps.

If your birds are outside, make your cages with rectangular cage wire; chain link and wrought-iron cages will allow rodents to enter. Outside cages that reach the floor should have cement floors, not dirt.

Predators

Birds that are outside are also fair game to birds of prey, raccoons, and other predators. One breeder we know has problems with birds of prey pulling on birds from outside the cages; another breeder almost lost a proven male Congo African Grey when a raccoon bit off the bird's foot as it tried to pull the bird through the cage (the bird survived).

In addition to these problems, other household pets and children can stress breeder pairs that are set up in backyards and family rooms. Many birds need privacy to breed.

Male Cockatoo Aggressiveness

One problem in breeding cockatoos is male aggression. Some males severely injure or kill females. This is especially true for Sulphur-crested Cockatoos; males have been known to rip off the females' beaks! We once set up a pair of Lesser Cockatoos; the male had been our pet for eight years. We monitored them for six months before putting the birds in a breeding cage. Four months later, the male attacked the female, who was on two eggs. He caused multiple fractures on one wing, and a compound fracture on the other. He also tore off a section of her upper and lower mandibles and fractured her skull. When we found her in the bottom of the cage, she seemed in good spirits despite her injuries. That evening, however, she died of internal injuries.

We also lost a proven female Moluccan Cockatoo in a similar situation. The male had broken her leg so badly that it had to be amputated by our veterinarian. She survived the surgery but died several hours later, probably of shock.

Male Cockatoo aggression does not occur in the majority of pairs set up for breeding, but it is a reality that anyone considering breeding Cockatoos should be aware of.

Important Notes:

- Male Cockatoos may suddenly become aggressive after years of successful breeding.
- Very tame ex-pet males have a greater chance of becoming aggressive with their females than wild, untamed birds.

- T-shaped nest boxes with two entrances will not prevent aggressive males from injuring females. Often a male may attack the female in the cage, not inside of the nest box. (We have used this type of box, and we did not like it because it was almost impossible to check on the eggs without the parents trampling them.)
- Clipping the males' wings and putting Cockatoo pairs in extra-large cages or flights may help.
- Once a male has severely attacked his mate, separate the pair permanently.

Working And Caring For Babies

Breeding exotic birds is very time consuming. It is impossible to work a full-time job and take care of an incubator full of eggs and a room full of babies. If you plan to breed only a few pairs of birds, it is not as bad, but you must still make arrangements for someone to feed your babies while you are at work, unless you have a very understanding boss who will let you take your babies to work with you.

Vacation, What's That?

Bird breeding is also very unpredictable. It is difficult to go on a vacation, because you never know exactly when your birds are going to be on eggs. If you do have an opportunity to get away, you must find someone capable of taking care of all of your birds. Many times when you finally get away, something goes wrong back home. It is best to plan your vacations when you have the least number of eggs and babies, so the chances of losing the aforementioned are lessened. Many of the larger species breed most of the year, except Amazon Parrots, which usually breed in the spring, which is always a busy time for most aviculturists.

When you are gone, your birds must be fed and watered every day, and it is best to have someone do this at the same time every morning, so they have fresh food available to them all day, and so the bird's feeding schedule is the same as usual.

Emergencies and Disasters

Living in Southern California, we are constantly coping with earthquakes and their effects on our birds. It is important to have an emergency plan in the event of a fire or any other natural disaster. In order to be prepared for an emergency you must:

1) Keep on hand enough carriers to accomodate your birds in case you need to evacuate your home. You can store inexpensive cardboard carriers flat and assemble them in seconds to transport your birds.
2) Purchase and store enough bottled drinking water in your home to sustain your family and your birds for one week.
3) If your aviaries are padlocked, make sure you have an extra key available for an emergency. Make sure someone in your family knows where it is!
4) If you have a lot of birds and/or an incubator full of eggs, you should purchase a gasoline-powered generator in case the electricity goes out. If you have only a few eggs to keep warm, wrap each egg in a tissue and hold it in your hand or next to your body until the power returns; human body temperature, 98.6° F. is ideal for keeping parrot eggs warm.
5) Never let your seed supply run completely out; try to have at least one week's supply on hand at all times.
6) Purchase and hang a fire extinguisher in each bird room and in other rooms in your home. (Not all types of fire extinguishers are safe to use around birds; CO_2 fire extinguishers can be dangerous.)
7) Make emergency plan guidelines and post them in a designate area; discuss the details with your family and helpers. Stage a mock disaster if you feel it would help better prepare you.
8) Keep first-aid supplies in a portable bag or box in case a bird injures itself.
9) Cycle your supplies so that all food, water and medicine is up to date. Check expiration dates regularly.
10) If your aviaries are outdoors and you live in an area that is subject to tropical storms and hurricanes, be sure that when the aviaries are built they are secure enough to withstand strong winds. In the event of a major storm, evacuate your birds, place any eggs in your

incubator, and wait out the storm. If you have time, tie any outside cages down with additional rope or heavy wire.

None of us like to think of something this unpleasant, but the more prepared we are for an emergency, the easier it will be to deal with if it happens.

Finally, breeding exotic birds can cause many heartaches and disappointments. You may lose a baby or an adult to a careless mistake like forgetting to check the temperature on your incubator after adjusting it or forgetting to lock the door on your breeders' cage. A bird can also get hurt if it gets out and climbs on another bird's cage. Sometimes birds die of mysterious causes or, even worse, a virus may develop in your aviary or nursery, killing many.

There is no way to really prepare yourself for everything. But the more you read and learn about birds, the more you will be able to handle even life-threatening situations.

CHAPTER 10

ILLNESS

The more observant you are, the sooner you will notice when one of your birds becomes ill. Babies can be weighed daily, but it is very difficult to check the weight of your breeder pairs when you don't handle them outside the cage; you can only go by how they look. By the time they *look* thin, they are usually dangerously thin.

Try to glance at the droppings in each cage every day when you feed. Bright yellow stools can indicate a liver problem, and red to dark-brown droppings could indicate blood in the stool. *Sometimes a dark brownish-black stool is indicative that your hen is going to lay an egg, however.*

When we set up our first pair of Scarlet Macaws, the male and female were both veterinarian checked and healthy. Six months later we noticed our hen had passed several large brownish-black stools and we were very alarmed because she had never done this before. She appeared fine, was eating and drinking and seemed to be at a good weight, but we called an avian veterinarian anyway. We told the veterinarian about the stools, and he suggested she could be hemorrhaging and said that we should bring her in right away.

We decided to call another avian veterinarian, who also breeds Macaws and other birds. He said, "Congratulations, you should have an egg in a day or two." Sure enough, we did. Still, it could have been a problem, but since we had had the pair checked six months before, we had not brought any new birds into our bird room, and because both looked good, we decided to chance not taking her in to the veterinarian. Now we're glad; if we had stressed her as the egg was forming, she could have become egg bound or she could have died.

Bacterial Infections

When you feed your birds, if you notice an unpleasant odor, one of your birds could have a bacterial infection. Smelly stools and thick greenish-brown stools indicate a harmful bacteria, such as *E.coli, Klebsiella, Proteus, Salmonella* or *Tuberculosis.*

Have an avian veterinarian culture the bird's cloaca (vent) and do a sensitivity test to see which antibiotic will best fight the bacteria. If it is a mild infection, you may be able to medicate the birds' water. Withhold all fruits and vegetables until you finish medicating (usually

10 to 14 days). Also, taste the water; if the medicine is strong tasting, the birds may not drink it. You may need to add some sweetener or juice to the water to hide the taste.

Psittacosis

If you have all of your birds tested in the beginning and keep them indoors, the chances of getting Psittacosis is fairly slim. (The tests are not 100 percent accurate, so a false negative is possible!) If one of your birds should break with Psittacosis, treat your whole collection. Tetracycline-coated pellets and millet are available. Caiques are finicky about the taste and may not eat treated pellets well. They do better on a cooked beans-and-corn mixture to which you can add children's cherry-flavored tetracycline.

Viruses

Among many viruses that cause problems, I will discuss the ones most bird breeders should become familiar with:

• **Pacheco's Virus** - This herpes virus is usually transmitted by birds that are stressed. Carriers are usually Conures, *but not always.* Death follows shortly after exposure. At this time, testing for Pacheco's Virus is usually done by liver biopsy. If a bird is thought to have or carry Pacheco's, it is usually treated with Acyclovir and vitamin injections. You must keep such a bird separate from the rest of your birds.

A vaccine against Pacheco's virus is now available but it does not guarantee immunity. We tried the vaccine but found it too stressful on our breeders.

• **Pox Virus** - Pox is carried from bird to bird by mosquitoes, mainly in tropical climates. Pox virus usually affects imported birds exposed to the virus, especially Blue-Fronted Amazons and Pionus Parrots.

There is a vaccine available for Pox, but it does not always prevent the virus. If you live in a tropical climate, you may wish to vaccinate; we do not use the vaccine on our breeders, because we do not live in a tropical area, and there are fewer mosquitoes in less humid climates.

Important Note: Both the vaccines for Pacheco's and Pox are oil-based injections; some birds develop large lumps in their chest muscles from the vaccine, so the vaccine must be administered subcutaneously (under the skin) rather than intramuscularly. It is important to realize that any psittacine could have an adverse physical reaction to a vaccine. Cockatoos seem to be more susceptible to this type of reaction; you may wish to consult with your veterinarian before vaccinating.

Papovavirus (Polyomavirus)

Papovavirus, now more specifically referred to as Polyomavirus, is carried by adult breeders and passed on to babies they feed. The babies are exposed to the virus every time they are fed by their parents. When the babies reach the pinfeather stage, the stress of all the new feather growth allows the virus to become active, and the baby bird may die quickly.

The most stressful stages of a baby's development are when it its pinfeathers start to open and when the bird weans to regular food. The parents remain carriers, so if they continue breeding, the cycle repeats itself. It is not known exactly how long infected parents should be prevented from breeding; however, it is recommended they go six months without a nest box to give the parents' immune systems time to overcome the virus.

If you have an outbreak of Papova in your nursery, and you have enough room in your home, you may wish to segregate each clutch of babies in a different room. If you have a lot of babies, try keeping similar birds in each room; for example, put all your African birds in one room, your South American birds in another, etc. This may increase your survival rate, if done early. Wear fresh rubber gloves and a clean lab coat in each room and feed in the same order every time: All sick birds should be handled last, regardless of cost.

Avian Serositis Virus

At the time of this writing, this virus is still in its "discovery" stage. First, you will notice that an affected baby appears bloated, especially under the abdomen. After a short time, the edema (bloating) is so severe that you can see and feel liquid under the skin. The bird appears droopy and listless and may breathe with difficulty due to the pressure on its heart and lungs.

We have seen this virus in Blue-headed Pionus and Apricot-headed Caiques we purchased at the ages of 2 to 4 weeks. Both species displayed bloating at around 5 to 6 weeks.

We followed an injectable-antibiotic therapy for both sets of birds, but once the bloating started, it was impossible to stop. Our veterinarian even tried periodically withdrawing the fluid from the Blue-headed Pionus' abdomen, but it didn't help. From the necropsy, we discovered that the liver of this bird was irreversibly damaged. When these birds appeared to be incurable and in distress, we chose euthanasia.

Macaw Wasting Disease

Macaw Wasting Disease can often be seen in imported birds under stress, especially Blue and Gold Macaws. It can be contagious to younger birds if an infected bird is shedding the virus in the same environment.

The first symptom is usually a noticeable amount of undigested, regurgitated seed in the cage. The infected bird may always seem hungry. The bird eats and eats, but fails to digest its food and starves to death.

Such birds are usually incurable. I do not know of a bird that has survived Macaw Wasting Disease. The proventriculus, a major organ of the digestive system, is usually severely affected. There have been several reported cases of Amazons and African Greys that contracted Wasting Disease; whether this is the same strain or something different has yet to be determined at this time.

Psittacine Beak and Feather Disease

Psittacine Beak and Feather Disease, or PBFD, is also known as Cockatoo Beak and Feather or Cockatoo Rot. This disease was orignally discovered in Cockatoos but can affect African Greys and other psittacines exposed to this disease. The disease destroys the immune system of the affected bird and robs it of its beautiful feathers and/or beak.

Ultimately, the bird succumbs to a secondary infection. The first symptoms are loss of powder, dirty feathers, broken, pinched off feathers with black tips, broken crest feathers that never seem to come in and appear pinched off, a chipped or overgrown beak or a soft beak.

The virus affecting these birds is extremely contagious and affected birds should be kept away from all other birds. The virus spreads in feather dander and bird dust. Birds over three years old seem to be less at risk than younger cockatoos.

At this time there is no cure, but a vaccine may soon be available to provide immunity to birds that are not infected with the virus. Experimentation is also being done with drugs in an attempt to reverse the effects of the disease in birds that have been infected with the virus.

Important Note: Avian veterinarians and aviculturists are experimenting with anti-viral drugs in hopes that they may find a way to halt outbreaks of Papova, Serositis, Macaw Wasting Disease, and PBFD. Some of these drugs are showing great promise when used alone and in combination with antibiotics (to prevent secondary infection).

Sarcocystis

Sarcocystis is a fatal disease caused by a protozoan parasite which is brought into the aviary by opossums. Cockatoos, Cockatiels and African Parrots are more susceptible to this disease, as are any psittacine collections in the Southern Florida areas. It may still occur in any area that opossums are present. Psittacines are sometimes affected by an intermediate host to the opossum, such as cockroaches, cowbirds and grackles. The cockroach may eat feces from an infected opossum and in turn be eaten by a parrot, which will also become infected. Cowbirds and grackles may eat from the ground where an infected opossum may have frequented and the droppings from the infected birds may infect an outdoor aviary.

Many large breeding facilities in Southern Florida have electric fences surrounding their farms in hopes to eradicate opossums near their birds. Unfortunately, this does not guarantee that the cockroaches or cowbirds or grackles will not visit, and bird breeders are constantly battling this problem in these areas. At this time there is no known prevention for this disease, and after infection, death is very quick.

CHAPTER 11

PROBLEMS WITH BABIES

There are so many things that can happen before and after you pull babies; it is *vital* that you pay attention to their progress.

Cold Babies

If you notice the babies are cold, pull them immediately. If they are under two weeks of age, put them in your incubator or brooder at 98° Fahrenheit and keep them there until they are strong enough for a brooder. If the babies are older, put them in a brooder at about 85°F.

Let the babies warm up before feeding them. The shock of warmed formula to their systems will be more stressful than delaying the feeding thirty minutes or so.

Babies that are too cool may have trouble digesting their formula at first because their systems have begun to shut down. Make the formula a little runny, and if the babies are lethargic or weak, add five percent of corn syrup (light or dark) or molasses to their next two or three feedings to boost their energy level.

Don't be quick to give up on cold babies left in the nest. We have rescued many of these from our aviaries and after they make it past the first 24 hours they are usually just fine.

Important Note: NEVER dispose of a cold baby that appears dead until you have attempted to revive it, and the baby has failed to show signs of life.

Hot Babies

In warmer climates, it often happens during a heat wave; in the nest, the babies become too hot and sometimes die. This happens more often with African Grey Parrots because they are such steadfast parents; they sit on the babies even under less than ideal conditions. This is also a very common problem with babies under *human* care. There is no good reason to lose a baby to improper temperature conditions! Many people do not like to use heating pads to keep babies warm because they have had a bad experience and burned a baby. We prefer to use a heating pad in a homemade brooder, following these precautions:

1) Do not use old heating pads; they may be unreliable. (Heating pads that are over five years old should be discarded.)
2) Never use a heating pad on Medium or High settings, use only Low settings. (Put a strong piece of cloth tape over the temperature switch so it cannot be accidentally changed.)
3) Always place the heating pad on the side and down 1/3 of the brooder unit, so the babies never stand directly on the heat source all the time.
4) Always keep the cover on the heating pad or wrap a towel around it.
5) Use layers of towels in between the heating pad and the box the babies are in to help protect the babies from direct contact with the pad.

If the babies get too hot, they will cry in distress. It is a different cry from a hunger cry. Overheated babies will also thrash around in their brooder. Sometimes there will be blood in the urine or feces (a pinkish stain) due to too much heat. As soon as the babies are moved to a cooler location, the urine will return to its normal color. The babies may also appear flushed and pant rapidly. Remove the babies from the box and place them on a soft cotton towel or T-shirt, until they seem comfortable. You can always tell when a clutch of babies is too hot; instead of grouping together in one heap, they separate to escape overheating.

Burned Feet

If you should inadvertently burn a baby's feet, arrange a new brooder and use cut out squares of T-shirts to line its nest. Soak the baby's feet in warm water with a few drops of Betadine® and apply Neosporin® or a similar antibiotic cream three times a day for about three days.

The soft T-shirt material is soothing to the baby's sore feet and the squares can be thrown away or washed to re-use. Be sure to change the baby frequently so it does not get too dirty.

For severe burns, consult your veterinarian.

Crop Burns

If you are diligent about stirring and checking the temperature of the food before you feed it to your babies, and you avoid using a microwave oven, you should not have any problems with burned crops.

Food that is heated or reheated in a microwave gets "hot spots," which, unless stirred thoroughly, can cause a burn. The temperature of the food should be between 100-104° Farenheit. We tend to feed food that is about 99-102°F. We triple check the formula by stirring thoroughly, drawing up the food in a syringe and testing at least 5 cc's on our wrist and then mouth.

You may purchase at any auto-parts supply store an air-conditioner thermometer, which can be used to stir the food. Never rely on a thermometer, however, always double check the food yourself.

If you do burn a baby's crop, you won't know it until a few days (and sometimes a week) later. The first sign of a crop burn is a bump or scab on the skin of the crop. Sometimes the skin will become black or blue.

Crop burns are more noticable on younger babies because there are fewer feathers over the crop skin area; on an older baby it is less noticable, because most or all of the crop is covered with feathers. Sometimes, a breeder or hand-feeder may not be aware that there is a a crop burn until the scab on the crop falls off, leaving a hole in the crop, out of which the formula spills out of when the bird is being fed. As soon as you notice anything irregular about a baby's crop, take it immediately to your avian veterinarian for examination and possible surgery. In a few cases, a mild burn (which looks like a raised pimple) may only require antibiotics as a preventative from a secondary infection. In most cases, however, the skin tissue is permanently damaged, and must be surgically removed, with the remaining healthy skin stretched and sewn together. To prevent the stitches from tearing open, you must feed the bird only tiny amounts of food, every hour or so, depending on the severity of the burn and the age of the bird.

After going through this experience once, most bird breeders learn their lesson and review their food-preparation habits. The nice thing about ready-mix hand-feeding formulas is that you can use hot water and add the formula to it; thus eliminating the need to heat up the food in the microwave.

Dirty Babies

A baby that collects dried formula or feces on its feet or body should be cleaned to prevent infection. If the baby is very young, a cotton swab dipped in warm water may be sufficient. For an older bird with feces on and around the vent, we usually use a warm wash-cloth to wipe off any material and blot the baby dry. Sometimes we carefully soak a really dirty baby in a shallow dish of warm water for several minutes; we make sure to keep the baby's head out of the water, and we work in a warm, draft-free room. After we have cleaned the baby off, we dry it with a warm towel. Older babies with feathers may require blow-drying; always use a low setting, never hot, and hold the blow dryer at least 12 inches away from the bird. It may take several baths to remove caked-on dirt.

Be sure to keep the baby's face and nostrils clean at all times. If dried food plugs a nostril, gently scrape it out with a clean fingernail or blunt object.

Important Note: Watch leg bands for build-up of dried feces. Clean any material off leg bands daily.

Poor Feeding Response and Lethargic Babies

A baby that is lethargic, does not cry and has a poor feeding response may be ill. Have it checked by your veterinarian and request a culture and sensitivity. Since the culture results won't be back for several days, ask your veterinarian if a mild, broad-spectrum antibiotic can be used meanwhile; it could save your baby's life.

When feeding a baby that doesn't bob, hold its beak between your thumb and forefinger. After you squirt a small bit of formula in the chick's mouth, gently bob its beak, head and neck up and down. If the baby is digesting on schedule, it will probably be okay.

Failure to Thrive

Anyone who has raised baby parrots for awhile has raised at least one baby like this: it eats and digests well, but it just won't gain weight. Always culture such a chick. Also check a birds with these signs for parasites; usually there will be nothing seriously wrong.

However, as the bird develops, it will become obvious that something is definitely wrong, the baby may suffer from Failure to Thrive Syndrome. Such a baby may grow slowly: the head may grow larger and soon appear out of proportion to its body. Although the eyes have not yet opened, they will appear to bulge out of the baby's head. This condition can be caused by a toxicity in the formula; by spoiled or old food; disinfectant residue on the syringe; a toxin in the environment; inappropriate use or abuse of antibiotics; malnutrition; or starvation due to lack of sufficient food.

It is very difficult to save a baby that does not thrive. The ones that live may be smaller than normal size as adults. You must be patient, persistent and give a lot of attention to these babies.

Slow Digestion and Sour Crop

Too often, if a baby digests its food slowly, its hand-feeder assumes the baby has sour crop. Because a baby digests slowly, however, does not necessarily mean it has sour crop. There are several possibilities. The formula could be too thick; the baby could be eating too much; or it could have a yeast infection such as *Candida*.

One sign of Sour Crop is a failure of the baby's crop to empty completely or at all by the next scheduled feeding. The baby may vomit some food from its crop, flinging its head from side to side. The baby's mouth and any vomited food will smell sour, like vinegar. The baby's eyes may be slightly dull or unresponsive because it feels very uncomfortable.

If the crop is very full, the old food must be diluted with warm distilled water and removed with a gavage tube. If you have never tube fed or gavaged a bird, have your veterinarian show you how so you will know how to do it. It is relatively easy to do once you learn how, but it is a technique that is very dangerous if done incorrectly.

If the baby is partially empty, you can warm up some water and add some baking soda (For 5 cc's or less, add a large pinch, for 10-20 cc's add about 1/8 tsp.) Carefully feed the water with baking soda, slowly so you don't aspirate the baby. The liquid should go through the baby quickly, and the next stools may be quite watery. Once the baby is completely empty, feed the formula watered down until you have the Sour Crop under control. If your chick continues to have problems with slow digestion, you should see your veterinarian about using Nystatin, or a similar antifungal that is used for treating yeast infections.

If there is no improvement in the baby after being on Nystatin for 24 hours, it probably has a bacterial infection, which requires additional medication.

We have noticed that strained spinach baby food in the jar speeds digestion in some babies, but it should not be used all of the time.

Overstretched Crop

Occasionally you may overfeed a baby. In an isolated incident, the baby will only take a little longer to empty. Just be sure to feed the baby less food when it does empty so it will be empty again at its regularly scheduled time. If you continue to overfeed, the skin of the crop will lose its elasticity and become overstretched. This is also referred to as Pendulous Crop. This condition is easy to reverse by making a simple "push-up bra" with stretchable gauze. (See photos.)

If this condition is not corrected, it can lead to more serious problems such as sour crop, bacterial infections or starvation, because the bird cannot properly digest its food. If you suspect your bird has an overstretched crop, do not continue to feed new food if the bird still has food in its crop. In such a case the old food will spoil in the crop. The bird will then become ill and be unable to digest any food. Your avian veterinarian would need to empty the affected bird's crop by using a gavage tube. A "Crop bra" should then be placed on the bird and left on until the crop skin regains its elasticity, usually about three to five days.

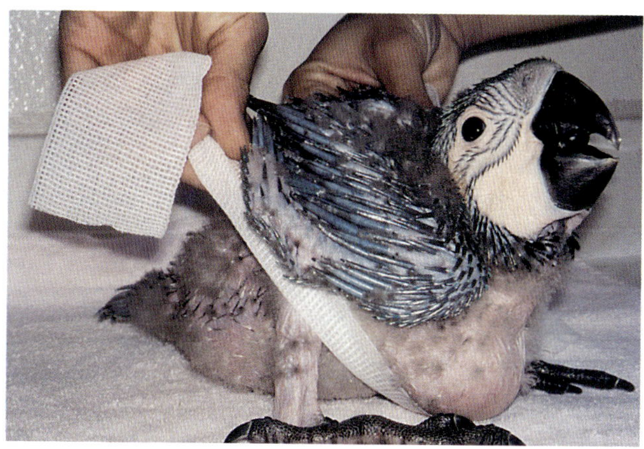

First, cut a piece of stretch gauze and wrap it under the baby's crop and behind the wings.

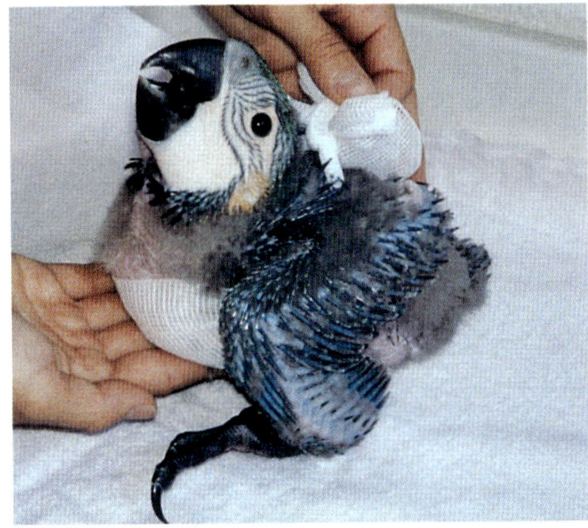

Next, pull the gauze so that it gently pulls the crop upward, and tie the ends of the gauze in a bow behind the wings. Check to make sure that the bra is not too tight, and readjust it if necessary.

Babies That Are Not Passing Stools

Healthy baby parrots pass stools several times in an hour. If your baby is not passing stools or passes them only infrequently, you should be concerned. This is especially common with very young and Day-1, incubator-hatched babies whose digestive systems are not developed.

Common causes:

- The baby is too hot and has become dehydrated.
- The baby is too cold, and its system has shut down.
- The formula is too thick and needs more liquid.
- The baby may have a bacterial infection.
- The baby has Failure to Thrive Syndrome.

The best thing to feed a baby that is not passing stools is an electrolyte replacer such as Lactated Ringers Solution or Pedialyte®. Feed one of these solutions undiluted to give your chick the full benefit of its ingredients.

Often a baby that does not pass regular stools is dehydrated or weak and the warm liquid stimulates elimination and re-hydrates the chick. Even if the chick is not dehydrated, these liquids will not hurt the baby. Just use care when you feed a solution such as these: a baby will choke more easily on a pure liquid than on formula. Feed any liquid slowly.

If you do not have any Ringers or Pedialyte® (available at most grocery stores), you may try adding a few drops of peanut oil, vegetable oil or coconut oil to the formula.

A baby that does not begin passing regular stools may have more serious medical problems.

Use of Electrolyte Solutions

Some breeders use electrolyte solutions such as Pedialyte® and Ringers the entire time the bird is hand-feeding. This is expensive and not necessary if the babies are healthy and digesting regularly.

Birds that are ill usually benefit from electrolyte solutions; however the salt and sugar in Ringers may not be good in some instances. It is best to consult with your veterinarian in each specific case.

Aspiration

As long as the bird you are feeding shows a good feeding response (or bobs), it is unlikely that you will aspirate it. If the baby does not feed easily, do not attempt to force-feed it; instead, try feeding with a spoon or more slowly with the syringe.

Always be sure to pull the syringe (or spoon) completely out of the bird's mouth when you are not feeding it formula, so the bird learns to breathe when it is not in the process of eating or swallowing.

You can usually tell if a bird is aspirated because it will cough, wheeze or sound different. A bird that is aspirated should be seen by an avian veterinarian, who may prescribe antibiotics to prevent a secondary infection. If the bird aspirates a small amount of food, it will usually cough most of it out over a period of several days. If there is a small amount of food in the lungs and the bird is otherwise healthy, the food mass will usually dissipate and the bird will be fine. If the bird has aspirated a large amount of food, it will usually develop pneumonia and die.

Baby's Eyes Not Opening

Sometimes one or both of the eyes have a little trouble opening and may appear to be sealed shut. In such a cases, moisten a cotton swab in warm water and gently roll it over the affected eye. If there is any swelling or discharge, have the eye checked for infection.

Important Note: When a baby's eyelids first slit open, its eyes are sensitive. Avoid excessive or close-up flash photography for about a week to avoid damaging the chick's retinas.

Baby Shakes or Shudders

Young babies may shake or shudder, especially after a feeding. African Grey babies often do this. If it is the first time you have fed babies, you may notice they are almost convulsive in their movements. This behavior usually disappears as the babies get older and more coordinated. It is believed to be a normal baby behavior in African Greys.

Baby's Neck is Twisted All Or Most Of The Time

A chick may sleep with its head tucked behind it, but if your baby seems to have its head tilted or neck twisted, you must have it examined by your veterinarian. The bird should be checked for a bacterial or respiratory infection and sometimes a "neck brace" may be recommended in addition to antibiotics.

It is very important that you try to arrange the baby's bed so the chick does not favor laying its head on one side. If you catch it early, all you need to do is prop the baby's head so it cannot tilt. You can help brace a young chick's neck with soft rolls of tissue after it has eaten.

Excessive head tilt problems can be permanent, so try all options while the bird is young and flexible. If you continue to have problems, you may need to review your techniques in incubation, hand-rearing and diet.

Broken Bones and Sprains

A baby bird's bones and muscles are easy to treat, and they

mend quickly. If you suspect your baby has a sprain or broken bone, take it to your veterinarian as soon as possible (within 24 hours) so the injury can be taped, splinted or cast in plaster. Ask your veterinarian to give you extra material and instructions on how to rewrap the injury in case the baby wiggles out of its first bandage. A baby grows so quickly it may need a new bandage every two days. By learning how to wrap, you can change a bandage if it gets dirty.

Curled Feet Or Improperly Aligned Toes

When a baby parrot first hatches, its feet may stay curled in a "fist" position for awhile. After two weeks the baby should appear more flat footed: Two toes should face forward, and two toes should face backward.

Sometimes three toes will point in the same direction, usually forward. When this happens, I take a small piece of tape and press it on the underside of the baby's foot, arranging the toes correctly. If the baby is older and so active that the tape falls off easily, you may cut a small piece of cardboard and tape with first-aid tape.

Spraddle Leg

Sometimes one or both of a baby's legs will stick out to the side. This condition is called Spraddle Leg. It can be caused by calcium deficiency, by lack of nesting material in the nest box or from the parent bird's weight.

Spraddle Leg, when caught *early*, can be corrected easily. It is extremely important to check your babies daily so you can pull a baby with Spraddle Leg.

First put the baby in a cone-shaped bowl or cup, with a small amount of paper or bedding. The angle of the bowl should force the baby's legs to sit underneath it. After about three to five days, the bird may be moved to a larger bowl until the correction is complete. In a severe case of Spraddle Leg the bird may need to be put in traction and have its legs taped together with a splint. The younger the bird, the easier the condition is to correct.

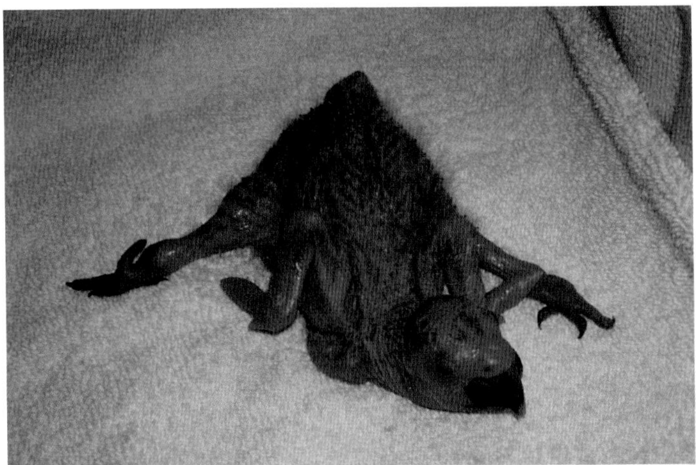

If a baby has one or both legs out to the side all of the time, it is called Spraddle Leg. This baby Macaw was pulled from the nest.

Curved Beaks

Rarely will smaller- to medium-size parrots have a problem with upper- or lower-mandible curvature. With larger parrots, especially Macaws, it is more common to see the upper mandible curve to the right or to the left. Most of the time this is due to the hand-feeder's technique. It is especially important to alternate sides if you experience this problem.

Another common problem is the hand feeding formula itself. We found that with one brand of commercially-prepared formula many of our Blue and Gold Macaws developed a slight curve in the upper mandible. When we changed formulas, the problem stopped. If the curve is limited to the tip of the upper mandible and the lower mandible is filed to keep it from growing higher on one side, the curvature will eventually grow out. If the upper mandible starts curving up high (close to the nostrils) then the bird will probably always have a curved beak.

If the upper and lower mandible are curved in opposite directions the condition is called scissor beak. This problem could be hereditary or the result of improper incubation techniques. This condition is irreversible and requires constant maintenance.

As soon as any beak abnormality is noticed, you must try to counteract the problem. Don't wait until the bird is eating solid foods to try something. Younger birds adapt to tape and other methods, and the amount of correction needed will be less when the bird is still growing, and the beak is still soft. One of our Hyacinth Macaws preferred to lay on one side all the time. We were also experimenting with commercial formulas at this time, and this combination resulted in a curvature in the upper mandible that started when the bird was 3 weeks old. If you looked at the bird, you could have seen that his beak curved to the right.

We wrapped plastic first-aid tape around the tip of the bird's beak in a clockwise direction. Then we pulled the end of the tape and attached it to his right "cheek." He could move his beak slightly and could be fed with the tape on. We were careful not to overfeed so he would not regurgitate. After about two or three feedings, the tape would become dirty or fall off, and we would wrap him up again. We did this for about a week. The results were very good, so we taped him for about another five days. At that time, the curve was very slight. By the time the bird weaned, you couldn't see a curve.

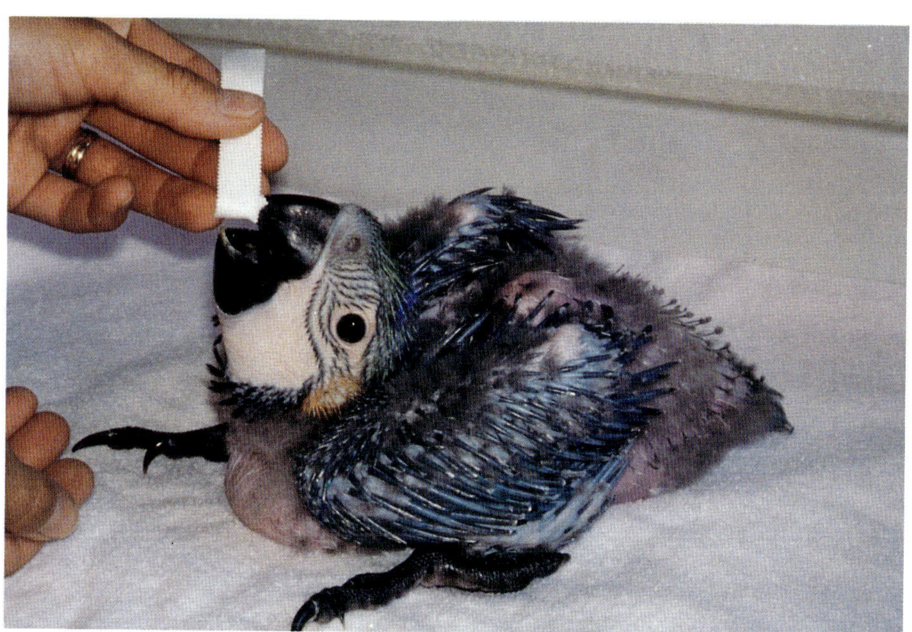

A curved beak can sometimes be corrected by taping it when the baby is young. First, you wrap the tape around the upper mandible several times.

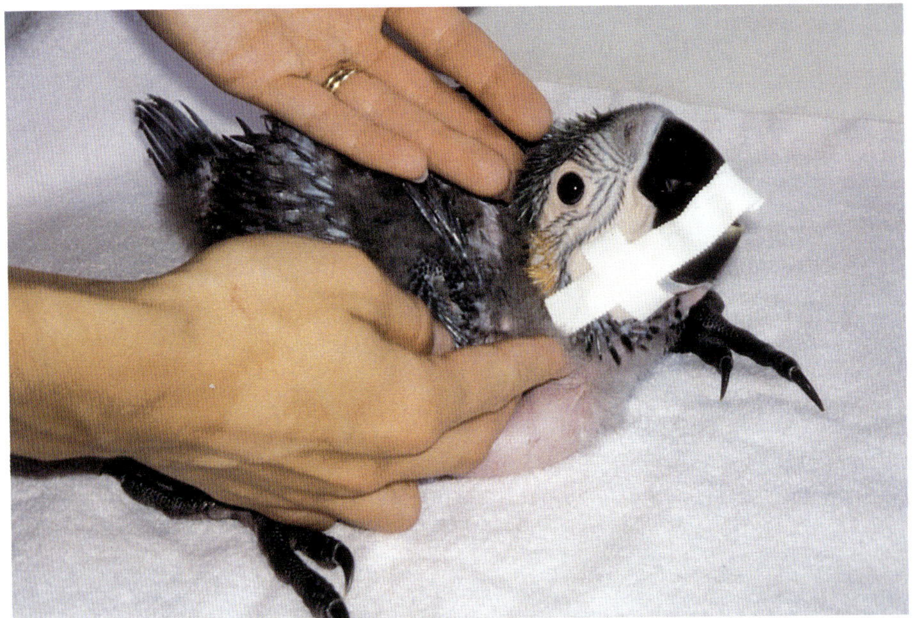

Next, you pull the tape over to the opposite side of the curvature and tape it to the lower mandible. Use a second piece of tape to secure the first piece. The tape should be fairly tight, but the bird should be able to slightly open and close its beak.

Cockatoos With The Upper Mandible Inside The Lower Mandible

Occasionally a baby cockatoo's upper mandible will curl inside the lower mandible. As soon as you notice this, you should try manipulating the tip of the beak with your fingers several times a day. Often this condition can be corrected later on when the beak is hard and can be filed with a dremel; sometimes it corrects itself. If you have more than one instance of a beak deformity, unless it is caused by the parents you must reconsider your incubation techniques (turning, temperature, humidity) and your hand-feeding technique and formula.

CONCLUSION

This book was written to help all of you who want to successfully breed larger parrots. It is important that you continue to read as much as possible about all different types of birds. Try to develop your medical skills so you can handle almost any situation. Always ask to watch your veterinarian when he examines and runs tests on your bird. Read medical journals and magazines on bird studies.

Remember also that you will need to spend time and money on your birds before they can reward you with healthy babies. Feed your birds the best foods you can offer and keep their breeding areas clean.

Always share your findings with others, and help those who are less experienced than you are.

Finally, I'd like to thank my husband, Omar, and all of you who helped and encouraged me to finish this book. A special thank you to our avian veterinarians for their support.

GLOSSARY

Blood Feather - A new feather that is nourished with blood until it emerges from its protective sheath.

Bonded Pair - A pair of birds that are compatible, but have not laid eggs.

Breeder - A bird that is used only for breeding purposes.

Clipped - A bird that has had the primary flight feathers clipped.

Close-banded - Domestic-bred birds banded with closed aluminum or steel bands with the breeder's initials and the year. (These closed bands should be placed above the toes and not on the leg. Bands should not be loose or tight.)

Clutch - Correct term to use when referring to the eggs or babies of a breeding pair.

Clutchmate - Sibling chicks in a nest.

Double Clutch - When a pair of birds goes to nest soon after the first clutch of eggs or babies are pulled.

Domestic Pair - A breeding pair in which both birds were domestically raised.

DNA Sexed - The sex of the bird was determined by using the DNA blood test. (There is usually a certificate from the lab.)

Egg bound - When a female bird cannot pass an egg.

Egg-laying Pair - A pair of birds has laid infertile eggs, but has not yet produced viable eggs or babies.

Ex-pet - A bird that was tame or hand-fed and was once kept as a pet.

F1 - A first generation, captive-bred bird.

F2 - A second generation, captive-bred bird.

Feather Picker - A bird that chews, plucks or mutilates its feathers from the neck down.

Full Flight - A bird that is not clipped.

Open band - Open bands made of stainless steel that are placed on birds that have cleared quarantine.

Pinioned - When a section of wing or bone is surgically removed so the bird cannot fly.

Proven Pair - When a pair of birds has mated, laid fertile eggs and produced babies.

Important Note: Buying a proven pair of birds does not guarantee babies. Unless the birds are set up properly and feel comfortable in their new location, they may stop breeding.

Proven Single Bird - A bird that has produced babies with another bird.

Seconds - Birds that are not perfect, i.e. missing toes, toenails, etc.

Sexed Bird or Sexed Pair - Birds that have been surgically sexed.

Tattooed - A mark under the wing for identification purposes.
(The "tattoo" is a small injection of black India ink under the skin.)

Wild-Caught Pair - A male and the female that were captured in the wild and imported.

PRODUCT INFORMATION

CARE-FRESH™ Animal Litter
Absorption Corp.
P.O. Box 5667
Bellingham, WA 98227
(800) 242-2287
FAX: (206) 671-8991

D-CA-FOS ® Calcium Supplement
NOLVASAN® Disinfectant
Fort Dodge Laboratories
Fort Dodge, IA 89103
(800) 634-6468

ENVIRO-CLEAN Disinfectant/Cleaner
Neon Pet Products, Inc.
P.O. Box 733
Cypress, CA 90630-0733
(714) 827-8471

DNA SEXING
Zoogen, Inc.
1105 Kennedy Place, Suite 4
Davis, CA 95616
(916) 756-8089

AVIA® VITAMINS
Nutra-Vet Research Corporation
201 Smith Street
Poughkeepsie, NY 12601
(914) 473-1900

NEKTON U.S.A.
14405 60th Street North
Clearwater, FL 34620
(813) 530-5000
FAX (813) 539-0647

HUMIDAIRE INCUBATOR COMPANY
217 West Wayne St.
New Madison, OH 45346
(513) 996-3001

GRUMBACH® INCUBATORS
c/o Swan Creek Supply
12240 Spencer Road
Saginaw, MI 48603
(517) 642-2716
FAX (517) 642-2130

LYON ELECTRIC COMPANY, INC. (Turn-X® and other Incubators)
2765 Main Street
Chula Vista, CA 92011
(619) 585-9900

DURO-LITE CORP. (Vitalites®)
Duro-Test Corp.
9 Law Drive
Fairfield, NJ 07007
(201) 808-1800

RUDY'S NEST BOXES
855 League St.
La Puente, CA 91744
(818) 918-2969

Breeder Records

Pair No. _____ Type of Birds: _____________________

Male Purchased on: ____/_____/_____

from: ____

Name of bird: _______________

___ Surgically Sexed

___ DNA Sexed

Date bird was sexed: _____/_____/_____ Veterinarian _______________

Vet checked with the following tests:

____

____

____

Brief History of bird:

____

____

____

Female Purchased on: ____/_____/_____

from: ____

Name of bird: _______________

___Surgically Sexed

___DNA Sexed

Date bird was sexed: _____/_____/_____ Veterinarian _______________

Vet checked with the following tests:

____

____

____

Brief History of bird:

____

____

____

Nesting Activity

Pair No. _____

List Date and Any Pertinent Information:

Egg Laid:	Egg Pulled:	Egg Hatched:	Babies Pulled:	Observations:

Weight Chart for Babies

Type of Bird: ________________ Parents (Pair No.) ____________

Hatch Date: _____________ __ Parent-fed __ Incubator Raised

Age:	Weight:	Gain or [Loss]	Additional Notes: